Answer Plans for the MRCGP

Answer Plans for the MRCGP

Julian Kilburn
MRCGP, FRCS (A&E), DA
Newcastle-upon-Tyne

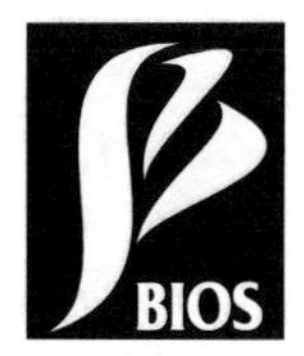

First published 2000

A CIP catalogue record for this book is available from the British Library.

ISBN 1 85996 144 4

BIOS Scientific Publishers Ltd
9 Newtec Place, Magdalen Road, Oxford OX4 1RE, UK
Tel. +44 (0)1865 726286. Fax. +44 (0)1865 246823
World Wide Web home page: http://www.bios.co.uk/

To my Mum, Dad and Robin Hood's Bay

Production Editor: Andrea Bosher.
Typeset by Marksbury Multimedia Ltd, Midsomer Norton, Bath, UK.
Printed by TJ International, Padstow, UK.

CONTENTS

Abbreviations vii
Preface ix

Part 1
Approach to essay questions 1
Practice questions 1–70 3

Part 2
The 4 minute vivas 109
Practice vivas with notes 111
Further viva practice 131
Further dilemmas 141

Appendices
Appendix 1: Appraising the article 143
Appendix 2: Key to evidence statements and grades of recommendations 145
Appendix 3: Hot topics 146

Index **147**

ABBREVIATIONS

A&E	accident and emergency
ACE	angiotensin converting enzyme
AIDS	acquired immune deficiency syndrome
BHS	British Hypertension Society
BJGP	British Journal of General Practice
BMA	British Medical Association
BMJ	British Medical Journal
BNF	British National Formulary
BP	blood pressure
BTS	British Thoracic Society
CHD	coronary heart disease
COPD	chronic obstructive pulmonary disease
CPN	community psychiatric nurse
DDS	deputizing doctor service
DRE	digital rectal examination
DTB	Drug and Therapeutics Bulletin
DU	duodenal ulcer
DVLA	Driver and Vehicle Licensing Agency
EBM	Evidence-Based Medicine journal
FBC	full blood count
FOB	faecal occult blood
GMC	General Medical Council
GMSC	General Medical Services Committee
GP	general practitioner
HDL	high-density lipoprotein
HIV	human immunodeficiency virus
HRT	hormone replacement therapy
ISIS	International Study of Infarct Survival
LDL	Low-density lipoprotein
LFT	Liver function tests
LMC	local medical committee
MCV	mean cellular volume
ME	myalgic encephalomyelitis
MI	myocardial infarction
MRC	Medical Research Council
MSU	mid-stream urine
NEJM	New England Journal of Medicine
NHS	National Health Service
NHSE	National Health Service Executive
NICE	National Institute of Clinical Excellence
NSAID	non-steroidal anti-inflammatory drugs
PACT	prescribing analysis and cost data
PCG	primary care groups
PHR	patient-held records
PID	pelvic inflammatory disease
PPI	proton-pump inhibitors

PSA	prostate-specific antigen
RCGP	Royal College of General Practitioners
RCT	randomized controlled trial
STD	sexually transmitted disease
TOP	termination of pregnancy
U&E	urea and electrolytes
UK	United Kingdom
UTI	urinary tract infection
WHO	World Health Organisation

PREFACE

The aim of taking an exam is to pass. I state this simply because it seems to be a point of controversy with those who become our teachers and examiners. They would have the aim of an exam to be to help you explore your inner thoughts and to prepare you for a career in general practice. This is a happy side effect because the exam is a good one. It forces you to think widely about issues and I am convinced that it made me a better doctor. Its emphasis on communication confirms this as *the* essential discipline of modern medical practice.

Also, because of the timing of the exam, the MRCGP is an extremely inconvenient exam to fail. A retake will almost certainly seriously interfere with your immediate future plans both personal and professional and involve a different topical syllabus to some extent.

That is enough about failure, the good news is that adequate preparation combined with a suitable awareness of the game has a high chance of success and it is in this area that the MRCGP candidate wins over those taking other postgraduate exams.

The MRCGP has changed again. The exam has become "modular", the written papers (MEQ and CRQ) have combined into one, and most importantly, the emphasis has been put heavily on the base of evidence. That means a lot of journal surfing for the worried candidate and many a happy hour at the photocopier.

The aim of this book is to present some of the newer formats of questions and answers from those resembling the old MEQ to those that request lists and summaries of evidence in a variety of styles. These are difficult to bluff and demand a familiarity with recent literature. However, they are likely to be on topics that are broadly familiar to you and the evidence you quote may well be "classic" evidence that takes many years to produce and has stood the test of time.

The book gives examples of questions and answers for the problem-solving and current awareness part of the new paper one. This makes up the major part of the paper. The critical appraisal techniques are well described in other texts.

In this guide, wherever possible, I have included references to relevant literature to encourage the necessity of always thinking of your answers as being supported by evidence as well as opinion.

The evidence that I have decided to quote in this section is mainly in the way of major guidelines, mega-trials and major reviews plus my own personal selection of papers. Secondary references are used including those from the journal *Evidence-Based Medicine* which appraises articles found in other journals. It is by no means exhaustive, but does point to some literature that you may find useful.

In this exam, any reference however vague is better than nothing. Do not be concerned with remembering the page numbers and third authors of articles. Where I have included them it is merely for you to be able to find should you so wish. It goes without saying that you will need to look up current references around the time you take the exam.

This author believes that the main hurdle in the exam is the written paper and this book is written for those who agree. Many of the available texts, although comprehensive on the issues of general practice, are of little use in the exam. At the time of writing no textbook of sample written questions and answers for the current requirements of the MRCGP exists. This is an attempt to put that right.

I have also included a section on the orals because of the difficulty in finding a useful guide and in getting useful practice. Initially I have suggested some points to consider including in your answer and then included some samples to think about and try out yourself.

Appendices have been included in a few key areas. The requirement to critically read a paper is always a concern. Using the guide at the back of the book and addressing a reasonable number of these points in your answer will turn a potential torment into a simple exercise (and easy marks), which anyone who keeps the checklist in their head can do.

A suggested list of hot topics is also included. Again this is a starting point rather than a definitive guide.

Julian Kilburn

APPROACH TO ESSAY QUESTIONS

The written test lasts 3.5 hours and the answers can be in short note format. You may be required to give answers by entering your response into a table or similar fixed format e.g. advantages on the left, disadvantages on the right. There are typically 12 questions, all of which must be answered, and they are weighted equally. Questions are marked by different examiners to a marking scheme. They require self-contained answers even if this involves repetition from a previous answer.

The answer plans here do not include critical appraisal or interpretation of written material presented to you. Examples of this can be found elsewhere or practised on any article with the aid of the guide at the back of this book.

The following questions and suggested answers focus on the identification of a relevant construct, or answer plan, for those questions which examine your ability to solve problems as well as integrate and apply evidence-based knowledge to current primary health care delivery in the United Kingdom.

A familiarity with current issues is expected.

- Read the recent journals, especially *BMJ* and *RCGP* of the last 2 years. Photocopy the front pages and list the topics you had not thought of or which are currently popular. Be aware of the Occasional Papers of the RCGP.
- Look through other sources such as *Evidence-Based Medicine* journal, *Drug and Therapeutics Bulletin, Prescribers' Journal*, other bulletins and magazines such as *Pulse, GP* and *The Practitioner* saving only interesting articles.

A quick search through electronic databases such as the Cochrane database can identify major systematic reviews.

THE ANSWER PLAN

Success in this part of the exam depends on your ability to come up with an appropriate construct for your answer. Spend a few minutes before starting to write thinking about the structure of your answer. There is no universal format and new questions in every exam may deliberately use a slightly different approach. Reading through the following pages will allow you to develop an adaptable style.

Always read the question and do as it asks. Questions that appear similar can require very different answers.

The following general format is suggested.

Introductory sentence and overview

This adds structure to the note form. A few statistics here can be a good start to creating an answer which *puts the problem into perspective.*

Construct: Five point plan

Try to identify five clear areas to develop. Use clear underlined headings. *This is how examiners will mark the papers. Your headings do not have to be the same as theirs but this approach will help you organize your answer and help the examiner to award you marks.*

Many but not all questions will fall into this plan and you must be prepared to be flexible. A mixture of question styles and answer formats are included in this book.

The use of evidence

Try and back up all your answers with some form of evidence if possible. Remember there are different levels of evidence (see Appendix 2).
The following is a suggestion of how you might use a hierarchy of evidence to help your responses.

Levels of hierarchy for exam purposes

1. Listing several high quality references to provide a balanced review of current opinion preferably using randomized controlled trials. Critical comment on the references will score highly
2. Summary of current thinking "recent evidence showed..."
3. Quotation of specific key paper "A BMJ review paper last year showed ..."
4. Recent news "a report recently suggested..."
5. Including some statistics in your answer

Be as specific as possible but come the exam any indication of reading is better than nothing. Rather than having the emphasis on quoting references, answers should reflect an understanding of the literature which allows a considered personal view.

The *Relevant Literature* sections include references in more than 80% of the questions. They vary from being extremely important in order to provide a balanced review of current opinion (and construct a good answer) to simply a useful reference for further reading in the same subject area. A maximum of four references are included here but your own reading will supplement these suggestions.

PRACTICE QUESTIONS 1–70

1. A 20 year old female patient of yours who admits to self-injecting with heroin has presented for the first time with a minor medical problem. What would you hope to achieve on this initial consultation?

INTRODUCTION

Harm minimization is seen by the Department of Health as a valuable and attainable goal and the next best thing to abstinence from drug misuse. The infinite potential of the consultation can be used to focus on this issue opportunistically.

FIVE POINT PLAN

Patient-related issues

- Explore patient's concerns and expectations about the presenting problem.
- Consider the possibility of underlying reasons for presenting.
- Explore patient's beliefs regarding own health.
- Assess motivation.

Issues for the doctor

- Proper management of the presenting problem.
- Good non-judgmental and opportunistic consultation technique to gain rapport and form relationship to build upon.
- Adequate history of drug problems and proper management of the presenting complaint.
- General health assessment including psychiatric morbidity.
- Generally supportive approach and further advice in specific areas.

Social/legal issues

- Explore social circumstances and evaluate home support.
- Ask about ongoing problems with the law.

Harm limitation

- Importance of sterile equipment.
- Information on needle-exchange programmes and injecting technique.
- Discuss blood-borne viral infection: HIV testing, hepatitis B vaccination.
- Sexual precautions.
- Patient's need to be conscious about own health needs.
- Invitation to contact health services.

Follow up plans

- Offer of help regarding review and/or substitute medication.
- Consideration of shared care referral to the local drug misuse team or specialist referral as necessary.

RELEVANT LITERATURE

Drug Misuse and Dependence – Guidelines on Clinical Management. Department of Health, 1999. (This consensus of the best available evidence is a guide for the GP)

Managing drug misusers, a guide. Kemp K. *The Practitioner.* **240:** 326–334, 1996.

Helping patients who misuse drugs. *DTB* **35(3)**, 1997.

2. Evaluate the evidence for treatment of hypertension in the elderly

INTRODUCTION

A great deal of evidence supports the value of identifying and treating hypertension in the elderly population. A widely accepted definition of hypertension is an average systolic blood pressure (BP) of >160 mmHg and/or a diastolic of 90 mmHg on three consecutive measurements. Isolated systolic hypertension is a sustained systolic BP >160 mmHg.

EVIDENCE

Non-pharmacological methods

Proven interventions include:

- salt reduction (may reduce BP by 6/4 mmHg);
- avoidance of excessive alcohol;
- increased exercise;
- stopping smoking;
- dietary changes.

(you may wish to quote evidence for each of these)

Pharmacological methods

EWPHE Trial 1985

The European Working Party on Hypertension in the Elderly studied 840 patients over 60 with thiazide medication vs. placebo and showed a significant reduction in cardiovascular mortality.

SHEP Trial 1991

The Systolic Hypertension in the Elderly Project studied 4736 patients with isolated systolic hypertension over 60. It compared thiazide to placebo showing a reduction in stroke, cardiovascular disease and mortality. Benefits extended to the over 80 age group with reductions in all major events. Treating isolated systolic hypertension reduces coronary heart disease (CHD) by 25%.

STOP-H Trial 1991

The Swedish Trial in Old Patients with Hypertension involved 1627 patients on thiazide, beta-blocker or placebo. It showed a significant reduction in fatal and non-fatal strokes, cardiovascular events and total morbidity.

MRC Trial 1992

This looked at 4396 patients aged 65–75 years and showed increased effectiveness of low-dose diuretic over beta-blockers with significantly reduced mortality in the diuretic group.

RELEVANT LITERATURE

The sympathetic approach to hypertension – a GP management guide. *Pulse*, 1999.

British Hypertension Society guidelines for hypertension management: summary. *BMJ* **319**: 630-635, 1999.

3. A 13 year old girl attends on her own complaining of a foul smelling vaginal discharge. What issues does this raise?

INTRODUCTION

This raises many issues of confidentiality, communication and consent.

FIVE POINT PLAN

Issues of consent

- Is there an acceptable level of understanding consistent with Gillick competence?
- Consent to prior sexual intercourse.
- Consent to examination.

Consultation

- The need to develop suitable rapport and take time.
- Difficulties of dealing with teenagers.
- The need for a third party, e.g. friend/parent/health visitor/chaperone.
- Consider relationship with parents.
- Adequate explanation of management/referral.
- Good record-keeping.

Sexual issues

- Establish sexual history.
- Risk of sexually transmitted disease.
- Possibility of pregnancy and contraceptive issues.
- Lifestyle advice.
- Sexual health of partner.

Legal and ethical issues

- Age of consent.
- Patient's right to confidentiality.
- There may be a possibility of rape or incest.
- Consider need to inform other agencies but also respect confidentiality.
- Awareness of possible action against doctor by parents.

Wider issues

- Making practice adolescent-friendly.
- Availability of local educational groups for adolescents.
- Follow-up and ongoing education.

RELEVANT LITERATURE

Gillick judgement 1985, *House of Lords*.

How can adolescents' health needs and concerns best be met? Jones R. *et al.* *BJGP* **47**: 631–4, 1997.

4. A 46 year old Muslim woman presents with a new problem. What are the issues for this consultation?

INTRODUCTION

Medicine in Britain is increasingly a transcultural specialty and it is essential that we try to understand the differences between us that impact on our ability to deliver primary care.

FIVE POINT PLAN

Issues for the patient

- Many believe in demand-led consultation and find appointments inconvenient.
- Eastern patients tend to be more frightened of pain and may need more reassurance.

Issues for the doctor

- Language barrier.
- Avoid transcultural conflict by showing awareness, courtesy, understanding and ability to negotiate.
- Show respect to a patient's culture, religion, ethnicity.
- Respect privacy and dignity (in accordance with the Patient's Charter).
- Assure the advocate that Eastern customs will be respected.
- May be using alternative medical practitioners with risk of drug interactions.

Family issues

- Illness in a family member can be a crisis for the whole family.
- Other members of family may wish to be seen in same appointment slot.

Role of the advocate

- Always accompanies Eastern women and may be relation or friend.
- Might speak for the patient making it a three-way consultation.
- Can act as chaperone, interpreter and mediator.
- May report back to the family on your treatment of the patient.

Practice issues

- Consider size of ethnic population.
- Discussion with colleagues to increase understanding in the practice.

RELEVANT LITERATURE

Cultural sensitivity in the consulting room. *GP.* February 1999

5. Lifestyle interventions are important in the secondary prevention of coronary heart disease. Evaluate this statement by listing the non-pharmacological factors on the left of the page and quoting evidence on the right.

INTRODUCTION

Coronary heart disease (CHD) is the single largest cause of death in the UK. Direct costs to the health service and social security were £1.6 billion in 1996. Indirect costs to the economy reach £10 billion each year.
There is now clear evidence of benefit of promoting lifestyle changes in patients who have already suffered myocardial infarction.

LIST AND EVIDENCE

Smoking cessation	• Reduces chance of further non-fatal MI and cardiovascular mortality by 50%. • Brief advice from the GP can achieve 5% cessation. • Reduces risk of post-infarction angina. • Advice plus nicotine replacement therapy is the best approach.
Diet	• Oily fish and fish oil supplements are high in polyunsaturates and reduce total mortality by 29% over 2 years. • Cardioprotective diet high in nuts, fruit and vegetables is better than a standard low-fat diet. • Evidence for vitamin supplements is incomplete.
Alcohol	• Moderate consumption of alcohol has protective effect reducing CHD by 30–35%. • Two to three glasses of wine per day is considered beneficial.
Obesity	• Inverse association between body mass index and long term risk of reinfarction. • Large waist circumference can identify those at increased cardiovascular risk.
Exercise/rehabilitation	• Cardiac rehabilitation reduces cardiovascular mortality from heart failure and further MI by 25%. • Reduces BP by 10/8 mmHg. • Moderately to vigorously active people have half the rate of MI of the physically inactive. • Aerobic exercise for 20–30 minutes three times a week is a suitable approach.
Stress reduction	• Part of most rehabilitation programmes but few firm data are available.

RELEVANT LITERATURE

Post-MI care in general practice. *Update,* April 1998.
Nurse-run clinics in primary care increased secondary prevention in coronary artery disease [RCT] Campbell, NC *et al. EBM* **4**(3), 1999.
The importance of diet and physical activity in the treatment of conditions managed in general practice. Little, P *et al., BJGP* **46:** 187–92, 1996.

6. A 55 year old man comes to your surgery after his recent admission for acute myocardial infarction (MI). You discover he has not taken any medication since his hospital prescription ran out 7 days after discharge. No discharge letter is available. List the groups of medicines for which he might be a candidate as a result of his MI. Discuss one piece of evidence which supports each answer.

INTRODUCTION

Pharmacological therapy is a valuable adjunct to lifestyle changes for secondary prevention of CHD. Many large well-conducted randomized placebo-controlled studies have illustrated this in recent years. This in association with local protocols should guide treatment.

LIST AND RELEVANT EVIDENCE

Antiplatelets
- The ISIS-2 trial proved the value of aspirin.
- It is the most cost-effective agent.
- 75–150 mg/day indefinitely for all patients unless contraindicated.
- Meta-analysis by the Antiplatelet Trialists Collaboration 1994 showed lower mortality and reinfarction rates by 25%.

Beta-blockers
- Continue indefinitely if not contraindicated e.g. respiratory disease, uncontrolled heart failure.
- Decreases mortality by 25% and reinfarction by 30%.
- Decreases sudden death (presumably an antiarrhythmic effect) by 30%.
- The ISIS-1 trial illustrated post-MI benefit.

ACE inhibitors
- Would probably benefit all post-MI patients.
- Greatest benefit in those with left ventricular dysfunction and those who develop early heart failure.
- SAVE trial (Survival and Ventricular Enlargement) – 2231 patients with low ejection fraction (<40%) but *no* signs of heart failure were given captopril post-MI. All-cause mortality was reduced by 19% over 3.5 years.
- The ISIS-4 trial also showed reduction in mortality post-MI.

Statins
- Cholesterol-lowering drugs in the form of HMG-CoA reductase inhibitors are effective post-MI and prescribing protocols are useful.
- The 4S Study (Scandinavian Simvastatin Survival Study Group) randomized 4444 patients with CHD aged 35–70 with total cholesterol 5.5–8.0 mmol/l and lowered LDL cholesterol by 35%. Follow up over 5.4 years showed a decrease in all cause mortality by 30% and coronary mortality by 42%.

RELEVANT LITERATURE

Post-MI Care (Pocket Guide) *Medical Imprint*, 1997.

7. Discuss your management of patients with heart failure referring to the literature.

INTRODUCTION

Heart failure represents 5% of all medical admissions and has a mortality rate higher than most cancers. Half of these patients will die within 5 years but treatments have been shown to be effective.

FIVE-POINT PLAN

Patient-centred issues

- Is there an understanding of the diagnosis?
- Explore concerns and worries.
- Assess lifestyle risk factors.

Issues for the doctor

- Gain common understanding with patient to help compliance with investigations and further treatment.
- Personal issues of keeping up to date with developments, awareness of practice or local protocols.

Issues for the primary care team

- Need to reinforce lifestyle messages.
- Team approach with nominated doctor and nurse.
- Development of specialist clinic to review risk factors.
- Consider training of specialist nurse or health visitor.
- Audit and development of guidelines/protocols.
- Use of a computer register.

Diagnosis and treatment

- Availability of open-access echocardiography.
- Lifestyle changes.
- Evidence-based prescribing and review.
- Follow-up with interview/blood tests.
- Consider need for shared care with hospital specialist.

Evidence base

Most of the evidence relates to the proven efficacy of ACE inhibitors in heart failure. They have been shown to reduce mortality and hospitalization.

SOLVD STUDY (*NEJM* 1992; **327:** 685–691).
The Studies of Left Ventricular Dysfunction prevention trial treated 4228 patients with an ejection fraction less than 30% using enalapril. Over 3 years mortality was reduced by 8% with a 20% reduction in hospital admission.

CONSENSUS (*NEJM* 1987; **316:** 1429–1435).
The Cooperative North Scandinavian Enalapril Survival Study showed a reduction in mortality of 40% at 6 months and 31% at 12 months in

patients with severe CCF. Treating 1000 people with moderate CCF saves 16 lives and 116 hospital admissions.

AIRE (*Lancet* 1993; **342:** 821–828).
Acute Infarction Ramipril Efficacy Study looked at 2006 patients in heart failure post-MI and showed a 26% reduction in mortality with ramipril at 15 months and an 11% absolute risk reduction over 3 years.

Unfortunately studies such as a BJGP review of heart failure in Liverpool also point to undertreatment in the community.

RELEVANT LITERATURE

Prevalence, aetiology and management of heart failure in general practice. Mair FS *et al. BJGP* **46:** 77–79, 1996.

8. List the benefits of exercise. Use evidence to support your claims.

INTRODUCTION

Physical exercise is effective for the primary, secondary and tertiary prevention of a wide range of physical and psychological illness. It is the commonest alterable lifestyle factor in cardiovascular disease but there is widespread lack of motivation. More evidence of benefit is required but clearly randomized trials are difficult to blind.

BENEFITS

Cardiovascular

- Reduces risk of stroke by 50% and heart attack by 60%.
- Meta-analysis has shown relative risk of 1.9 in sedentary occupations.
- Reduces risk of coronary death post-MI by 25%.
- Reduces blood pressure by 10/8 mmHg.
- Increases HDL-cholesterol.

Skeletal

- Weight-bearing exercise increases bone density at all ages.
- Can reduce risk of hip fracture by at least half.
- Improves flexibility in joints and reduces accidents.

Psychiatric

- Relieves anxiety.
- Has proven direct anti-depressant effect.

Other areas

- Improved glucose tolerance in non-insulin-dependent diabetes.
- Reduces handicap in the disabled.
- Reduced incidence of breast and colon cancer in the physically active.
- Increases longevity.

RELEVANT LITERATURE

Total physical activity and mortality were inversely related in men. Lee, IM *et al.*, *EBM* Nov/Dec, 1995.
Benefits of exercise in health and disease. Fentem, PH, *BMJ* **308:** 1291–1295.
Strategies for prevention of osteoporosis and hip fracture. Law, *et al.*, *BMJ* **303:** 453–459, 1991.

9. List the factors which influence a general practitioner's prescribing.

INTRODUCTION

The decision to prescribe is multifactorial. Evidence indicates that resistance to change is overcome by different types of cue in different doctors. This implies that individual styles require separate targeting and that traditional education cannot therefore be expected to bring about large scale change.

LIST

Intrinsic influences on the doctor

- Personal background and clinical experience with own patients in practice.
- Doctor's personality and keenness to try new treatments.
- Doctor's expectations of patients' likely compliance with medication.
- Doctor's perception of patients' expectations: one study showed if the GP felt the patient expected medication then they were ten times likelier to receive it. Patients who actually expected medication were only three times likelier to receive it.
- Willingness to prescribe without seeing patient.
- Experience of using a drug yourself and forming a personal opinion of effectiveness and side effects.
- Challenge from a specific event e.g. death of patient from amitriptyline overdose.
- Reinforcement of actions by positive feedback from patients.
- Inhibition of behaviour with negative reports and anecdotal evidence of adverse effects.
- Knowledge and clear understanding of the actions of a new drug.
- Use of prescription to close difficult and time-consuming consultations.
- Accumulation of pressure from clinical material e.g. articles, lectures (slow adaptive process).

Extrinsic influences

- Effect of the media on awareness and attitudes towards drugs in doctors and patients.
- PACT data.
- Local and central pressures of cost.
- Organization of practice and availability and accessibility of partners.
- Opinions of trusted colleagues e.g. partner or local consultant.
- Direct exposure to new methods of practice e.g. appointment of recently-trained partner.
- Contact with other colleagues e.g. feedback from pharmacist regarding your prescriptions.
- Contact with and opinion of pharmaceutical representatives.

RELEVANT LITERATURE

A study of general practitioners' reasons for changing their prescribing behaviour. Armstrong, D. *et al., BMJ* **312:** 949–952, 1996.

Factors which influence the decision whether or not to prescribe: the dilemma facing general practitioners. Bradley, CP. *BJGP* **42:**454–458, 1992.

Prescribing behaviour in clinical practice: patient's expectations and doctors' perceptions of patient expectations. Cockburn, J. *BMJ* **315:**520–523, 1997.

10. The last extra in your surgery admits to being addicted to heroin and requests your help coming off it. You do not have a fixed practice policy. What problems does this present?

INTRODUCTION

Around 30 000 drug misusers are identified on the Regional Drug Misuse Databases (1998 figures). The balance is towards males in the ratio 3:1 and over half report heroin as their major drug of misuse. The average GP may see one to two new cases per year. In recent years there has been a marked increase in drug-related deaths in the 15–19 year age group.

FIVE POINT PLAN

Issues for the doctor–patient relationship

- Minimum requirement is provision of care for all general health needs and drug related problems.
- Concerns over personal ability, experience and need for training in dealing with drug misuse.
- Awareness of range of local services.
- Development of mutual trust.

Issues for the practice

- Demonstration of competence in dealing with drug misusers.
- Development of agreed practice policy.
- Increased workload with demanding patients.
- Consider designating time and resources.
- Possibility of extra funding from the health authority.
- Danger of staff harassment.

Treatment issues

- Importance of contact with treatment services.
- Advice on risk reduction behaviour to limit harm.
- Supportive role.
- Problems with prescription of substitute medication.
- Comprehensive record keeping.
- Follow-up regime.

Legal issues

- Notification to the Home Office Addicts Index is advisable but no longer compulsory.
- High rate of criminal behaviour amongst users.
- Special licence from Home Office required for diamorphine prescription.

Role of other professions

- Agreement with pharmacists for supervision of substitute medication.

- Role of specialist services, community drug teams and guidelines for referral.
- Shared care model utilizes different skills most effectively.

RELEVANT LITERATURE

Drug Misuse and Dependence – Guidelines on Clinical Management. *Dept. of Health* 1999.

11. You are asked to co-ordinate the diabetic care in your practice. Describe your approach mentioning the evidence you would employ.

INTRODUCTION

New evidence supports the need for tight glycaemic control and intensive blood pressure management in diabetics. This is best provided by a well-structured framework in primary care.

FIVE POINT PLAN

Define objectives and priorities

- Select baseline and follow-up parameters.
- Aim for tight glucose control and good hypertensive control.
- Evidence has shown that intensive glycaemic control improves risk for many diabetic complications in Type I and 2 diabetes especially microvascular pathology.

Define methods

- Opportunistic or clinic based i.e. normal surgery time, mini-clinics or "diabetic days".
- Discuss methods of case-finding e.g. computerized prescriptions.
- Consistent treatment based on accepted evidence.
- Effective recall system.
- Meetings with relevant parties e.g. local diabetologist and community diabetic nurse.
- Evidence has shown that lack of structure of diabetic care is associated with increased mortality but organizational improvements can achieve equivalent standards or better.

Issues for the primary care team

- The need for a motivated team.
- Nominated doctor and nurse as well as chiropodist/optometrist.
- Develop evidence-based practice guidelines.
- Consider skill mix.
- Use of the shared care model.

Practice management issues

- Resources and financing.
- Need for special equipment.
- Accurate register of diabetic patients e.g. note tagging.
- Inclusion in practice brochure.

Review and audit

- Failsafes e.g. review of defaulters.
- Regular assessment of percentage seen and level of control.
- Audit e.g. reductions in out of hours calls, patient's views.

RELEVANT LITERATURE

UKPDS. The UK Prospective Diabetic Study involved >5000 people with Type 2 diabetes in 23 centres over 10 years. (*BMJ* **317:** 703–720, 1998).

DCCT. The Diabetes Control and Complications Trial 1993 (*NEJM* **329:** 977–986, 1993).

Diabetes care in general practice: meta-analysis of randomised controlled trials. Griffin, S. *BMJ* **317:** 390–396, 1998.

Implications of the UKPDS for general practice care of type 2 diabetes. Kinmonth, AL *et al.* *BJGP* **49:** 692–694, 1999.

12. Antihypertensive therapy is of particular importance in type 2 diabetics. What evidence is there to support this statement?

INTRODUCTION

Hypertension is a common and potent risk factor for vascular disease in diabetics. The clinical benefits of aggressive treatment are now clear.

GENERAL EVIDENCE

- Hypertension frequently precedes the diagnosis of diabetes.
- Studies have shown that 43% of men and 52% of women with type 2 diabetes are hypertensive.
- Hypertension and diabetes have similar risk factors e.g. obesity.
- The risk of coronary heart disease is increased by 2–3 times and for heart failure is five times normal.
- Non-pharmacological interventions are useful but most will need medication.
- Target BP is now 140/80 mmHg.

SPECIFIC EVIDENCE – FOUR MAJOR TRIALS

These trials used a variety of agents to achieve BP control.

UKPDS – The UK Prospective Diabetes Study 1998

($n = 1148$ all diabetic)
The major embedded study in this group was related to BP control. Follow up over 8 years showed that a modest reduction in BP by 10/5 mmHg was associated with major reduction in risks of diabetes-related death, stroke, heart failure and microvascular and diabetic endpoints. There may even be more benefit from control of BP than of glucose.

SHEP – The Systolic Hypertension in the Elderly Programme 1991

($n = 4736$, diabetic $n = 583$)
This showed greater benefit of hypertensive control in the diabetic sub-group.

HOT – Hypertension Optimal Treatment trial

($n = 18\,790$, diabetic $n = 1501$)
The Hypertension Optimal Treatment trial showed up to 50% reduction in major cardiovascular events with a diastolic target of 80 mmHg in diabetics.

SYST-EUR 1997

($n = 4695$, diabetic $n = 492$)
This found that with adequate antihypertensive treatment the excess risk of diabetes was almost completely eliminated with approximately 70% reductions in cardiovascular mortality and all cardiovascular end-points.

RELEVANT LITERATURE

The sympathetic approach to hypertension – a GP management guide. *Pulse* 1999.

Type 2 diabetes and the UKPDS. Greenwood R., *GP* January 15, 1999.

13. One of your partners has given your next patient temazepam for the last month and written "Do not repeat" in the notes. The patient attends saying he cannot sleep without it. What issues does this raise?

INTRODUCTION

Insomnia is a common symptom with four out of ten people complaining they often sleep poorly.

FIVE POINT PLAN

Issues for the patient

- Underlying reasons for presentation.
- Expectations of prescription.
- Social circumstances and associated problems e.g. stress, bereavement.

Issues for the doctor

- The need for good communication skills and a sympathetic approach.
- Irritation at lack of substantial symptoms.
- Is there underlying physical or mental illness?
- Perceived pressure to prescribe.

Issues for the practice

- Prescribing policy for hypnotics and the need for a practice formulary.
- Availability of behavioural and cognitive therapy locally.
- Structured programme for identifying benzodiazepine users.

Treatment issues

- Non-drug versus drug management.
- Availability of further help e.g. relaxation guides.
- Advice on "sleep hygiene".
- Need for follow-up.

Wider issues

- The prescription of addictive benzodiazepines.
- The Committee on the Safety of Medicines only recommends treatment with benzodiazepines for insomnia if severe and disabling and if possible only intermittently.

RELEVANT LITERATURE

Management of anxiety and insomnia. *MeReC Bulletin* **6(10),** 1995.
Insomnia (Pocket Guide) *Medical Imprint*, 1997.

14. Discuss the recent evidence regarding the use of lipid-lowering drugs in general practice.

INTRODUCTION

Cholesterol levels have a well-documented relationship with coronary heart disease largely mediated through LDL-cholesterol with a protective effect from HDL-cholesterol. Numerous trials have shown the benefit of lower serum cholesterol.

EVIDENCE

Primary prevention

WOSCOPS

The West of Scotland Primary Prevention Study involved 6595 males aged 45–65 with total cholesterol above 6.5 mmol/l but no history of CHD on pravastatin or placebo for 5 years.
Key results: Total cholesterol decreased by 20% and death from coronary heart disease by 31%. Treating 1000 patients for 5 years would prevent seven deaths from CHD and 20 non-fatal MIs.

Secondary prevention

The 4S trial

The Scandinavian Simvastatin Survival Study involved 4444 patients (81% were men) between 35 and 70 years with known ischaemic heart disease and total cholesterol of 5.5–8.0 mmol/l and used simvastatin or placebo over 5.4 years.
Key results: A lowering of total cholesterol by 20% was produced. This gave a reduction in total mortality by 30% and in coronary deaths by 42%.

CARE

The Cholesterol and Recurrent Event study involved 4000 post-MI patients under 75 with a total cholesterol under 6.2 mmol/l. It was a 5 year study of pravastatin versus placebo.
Key results: Secondary prevention was shown to be effective in lower ranges of cholesterol. Fatal and non-fatal cardiac events were reduced by 24% and stroke by 31%.

Cost-effectiveness

Secondary prevention has been shown to be highly cost-effective at £5000 per life year saved with simvastatin.
As costs fall thresholds for intervention may fall.
Targeting treatment where there is less than a 2% per annum risk may not be cost-effective.

Management guidance
Tables recommended by the Standing Medical Advisory Committee (London) take into account other risk factors and produce guidance on when to treat.

RELEVANT LITERATURE

Hyperlipidaemia (Pocket Guide), *Medical Imprint*, 1998.
Management of hyperlipidaemia. *DTB*, **34(12),** 1996.

15. Mrs Jones is 48 years old and asymptomatic. You know that her mother who suffered from angina died 2 years ago following a fractured femur. She asks you if you feel she needs to be on hormone replacement therapy. How would you manage this consultation?

INTRODUCTION

There are different perceptions of the benefits of HRT. Recent evidence has helped to evaluate the risks and benefits.

FIVE POINT PLAN

Issues for the patient

- Explore beliefs and understanding regarding HRT.
- Alteration in social circumstances or relationships.
- Understand patient's expectations.
- Ongoing need for contraception.

Issues for the doctor

- Need for empathetic consultation skills.
- Awareness of current evidence.
- Opportunistic health promotion e.g. maintenance of physical activity and adequate calcium intake, BP, smoking, lifestyle.
- Time-consuming consultation.
- Medical assessment.

Issues for the practice

- Use of clinical nurse for education and monitoring/lifestyle risk factors.
- Use of written material, videos etc.
- Practice formulary for prescribing.
- Prescribing costs.

Information-sharing

- Explanation of physiology.
- Risk–benefit assessment.
- Awareness of side-effects.
- Mutual agreement on different therapeutic approaches.
- Importance of the patient's own unhurried choice.

Evidence-based decision re principal effects

Cardiac

There may be a protective role in patients with significant risk factors or previous MI but clear evidence inconclusive. Trials have been short and have suffered from selection bias. The first RCT of women with known heart disease showed no overall benefit in a period of 4 years despite improvement in lipid profiles.

Osteoporosis

There is a clear role in proven osteoporosis and those at increased risk when HRT increases bone density in hip and spine. It can almost half the risk of osteoporotic fracture if given for 5–10 years but benefit is quickly lost on stopping. Likely to be more cost-effective when older e.g. >65 years but controversy persists on when best to start.

These potential benefits to be weighed against the downside which includes a small increase in breast malignancy and endometrial cancer.

RELEVANT LITERATURE

Hormone Replacement Therapy [Review] Barret-Connor E., *BMJ* **317:** 457–461, 1998.

HRT may not prevent against further heart attacks. Berger A, *BMJ* **317:** 556, 1998.

Hormone Replacement Therapy, *DTB,* **34(11),** 1996.

16. A 28 year old married businesswoman who is planning a pregnancy asks for advice before becoming pregnant. What areas do you wish to cover?

INTRODUCTION

Few women ask for advice before conception despite the fact that the prebirth environment can affect the development of significant medical problems such as cardiovascular and respiratory disease. Less than 40% of women in one study thought preconceptual counselling was essential and more than 10% thought it of no importance. The best way to give advice is probably opportunistically to women of child-bearing age.

FIVE POINT PLAN

Smoking

- Many continue to smoke but advice against it has been shown to be useful.
- Increased risk of spontaneous abortion, preterm delivery, low birth weight and perinatal death.

Need for supplements

- Folate supplements at 0.4 mg/day needed from before conception till week 12 of pregnancy for primary prevention of neural tube defect (5 mg per day if previously affected pregnancy).
- One recent study showed that less than 50% of women knew to take folic acid, despite awareness improving in recent years.
- No need for routine iron supplement; measure haemoglobin if suspected.

Diet

- Avoidance of slimming and aim for appropriate weight gain.
- Avoidance of excessive vitamin A e.g. liver products.
- Low alcohol intake.

Work

- Advice on statutory maternity pay.
- Advice on risk avoidance in the workplace.

Risk of infections

- Checking of rubella antibodies and need for vaccination.
- Assess previous history of chickenpox and discuss controversial area of varicella vaccine or need for zoster immunoglobulin in case of recent contact.

RELEVANT LITERATURE

Routine iron supplements in pregnancy are unnecessary. *DTB* **32:** 30–31, 1994.

Folic acid and neural tube defects: guidelines on prevention. *Dept of Health*, 1992.

Preconception care: who needs it, who wants it and how should it be provided. Wallace M, *et al.*, 1998 **48:** 963–6, *BJGP*.

17. Your practice is situated in a traditional mining area and you see many chronic bronchitics. What factors are important in delivering care to this group of patients?

INTRODUCTION

COPD is a chronic, slowly progressive disease responsible for a great deal of morbidity and mortality. In over 90% of cases the sole cause is smoking.

FIVE POINT PLAN

Issues for the patient

- Limitation of lifestyle.
- Loss of autonomy.
- Beliefs regarding the dangers of smoking.

Issues for the doctor

- Awareness of British Thoracic Society guidelines.
- Knowledge of local services.
- Need for shared care.

Issues for the practice

- Purchase of a spirometer or availability of open-access spirometry.
- The need for screening.
- Costs of time and burden to the practice of extra workload compared to savings on inappropriate medication.
- Need for extra staff training to ensure reliability.
- Need for development of practice guidelines.
- Influenza vaccination programme.

Clinical management plan

- Health promotion emphasizing smoking cessation.
- Rational use of medications to alleviate symptoms.
- Checking inhaler technique.
- Coping with limitations and modification of environment.

Evidence issues

- Discuss the use of spirometry mentioning the reproducibility and specificity of FEV1 as a diagnostic tool as well as a measure of the severity of disease. It is the only method for early diagnosis in asymptomatic individuals. Compare this to the potentially misleading peak expiratory flow.
- Decreasing acceptability of a purely clinical diagnosis.
- Value of smoking cessation.

RELEVANT LITERATURE

The management of COPD. *MeReC Bulletin* **9(10),** 1998.
BTS guidelines for the management of COPD. *Thorax* **52(suppl 5),** 1997.
Peak flow meters and spirometers in general practice. *DTB* **35(7),** 1997.

18. While visiting an elderly patient you are taken aside by her friend who shows you a cupboard full of several month's prescriptions. It appears she has collected the medicines but does not take the treatment. What issues does this raise?

INTRODUCTION

Non-compliance (or non-adherence) is estimated to contribute to therapeutic failure in about 50% of prescribing. It represents a major clinical challenge.

THREE POINT PLAN

(Although you could devise a five point plan for this answer, the subject falls nicely into three major areas)

Causative factors

Many of these stem from a breakdown in communication.

- Drug regime: doses too frequent or complicated by too many drugs.
- Unacceptable side-effects.
- Quality of the doctor–patient relationship e.g. patient not sufficiently involved in management decision.
- Incomplete advice from doctor about medicines.
- Patent's beliefs about illness.
- Unable to afford prescriptions.
- Cultural or religious barriers.
- Older, more vulnerable patients more likely to hoard medications.

Consequences of non-compliance

- Loss of benefit e.g. antihypertensive prevention strategies for heart disease and strokes undermined.
- Unnecessary additional drugs prescribed.
- Unexpected withdrawal effects.
- Resources wasted.

Ways of aiding compliance

- Methods of checking compliance (questioning, counting tablets, measuring blood levels).
- Communication issues: consider patient's ideas, concerns and expectations.
- Use simple language and repeat important messages.
- Tailor prescribing e.g. modified-release drugs, shorter courses of treatment.
- Clear verbal and written instructions.
- Review and feedback to patient (and relatives if available).

RELEVANT LITERATURE

Compliance with drug therapy. *Prescriber's Journal* **39**(1), 1999

Primary non-compliance with prescribed medication in primary care. Beardon P, *et al. BMJ* **307:** 846–848, 1993.

Why treatments fail. *DTB* **36(6),** 1998.

19. With reference to recent evidence, discuss the practical ways in which the primary care team can assist patients to stop smoking?

INTRODUCTION

Smoking is the most important avoidable health hazard. Smokers have a 50% chance of dying from their habit.
Seventy per cent of smokers want to give up smoking.

FIVE POINT PLAN

Issues for the patient

- Awareness of the dangers.
- Personal motivation to stop.
- Peer pressure.

Issues for the doctor

- Time constraints.
- Awareness of the evidence.
- Assessment of patient's motivation.
- Encouragement and advice on benefits of stopping.
- Assess dependency and need for nicotine replacement (is a cigarette needed within half an hour of waking?).

Helping smokers stop

- The four A's – Ask (about smoking at every opportunity), Advise (all smokers to stop), Assist (the smoker to stop), Arrange (follow-up).
- Specific techniques – naming the day, removing temptation, enlisting help etc.

Evidence issues

Of the proven benefits of stopping

- Excess risk of serious heart disease halved within a year of stopping.
- Risk of stroke halves by 5 years following cessation.
- Ex-smokers can restore their life expectancy to that of non-smokers.

Of cost-effectiveness

- Brief advice at each encounter is highly cost-effective.
- Use of transdermal nicotine patches over 12 weeks is very cost-effective and doubles the success rate of brief advice at 1 year.

Wider issues

- Prescribing: availability of nicotine replacement therapy on the NHS.
- Resources: up to £60 million pounds available for NHS smoking cessation initiatives.
- Government action: ban on tobacco sponsorship of sports (due to be fully phased out by 2006). Consideration of further legislation related to smoking in public.

RELEVANT LITERATURE

Smoking Kills – a white paper on tobacco. *Dept of Health,* Dec. 1998.

Simple support and nicotine replacement improve quit rates for smokers (review of RCTs). Law M, Tang JL, *EBM* Mar/Apr 1996: 91.

Smoking Cessation: evidence-based recommendations for the healthcare system. Raw M *et al., BMJ* **318:** 182–185, 1999.

20. The senior partner in your practice asks you to review the way you provide for the health needs of children. What are the issues of providing a good standard of care to children and adolescents?

INTRODUCTION

The GP provides the first and most frequent point of contact with healthcare for children. A comprehensive service should involve special consideration of their changing needs.

FIVE POINT PLAN

Communication

- Gaining rapport with adolescents to identify needs.
- Awareness of hidden agendas.
- Issues of communicating with parents.

Practice organization

- Specified children's waiting area (toys, changing area).
- Safety issues for the building.
- Written policies on seeing children.
- Open-access clinics for teenagers.
- Posters advertising policies in the practice.
- Nominated doctor to review policies towards minors.
- Clear policies on how to contact practice in emergency.
- Medical surveillance checks and vaccination programmes.

The Primary Care Team

- Need to keep up to date with changing needs.
- Involvement of health visitor input/school nurse.
- Provision of sexual health advice.

Confidentiality

- The duty of confidentiality to a minor is as great as that to any other individual.
- Many adolescents fear a break in confidence from their GP.

Legal framework

- The Gillick judgement 1985 by the House of Lords accepted that the right of a parent to decide a child's treatment terminates when a child achieves sufficient understanding and intelligence to enable an understanding of what is proposed.
- The well being of the child is paramount: a child at risk of "significant harm" necessitates reporting to the social services.

RELEVANT LITERATURE

Child Friendly Primary Health Care. Hogg C. *Action for Sick Children*, 1998.
"Health for All Children" and the new contract (editorial) Hall DMB, *BMJ* **299:** 1352–1353, 1989.

21. A 19 year old girl attends with her boyfriend requesting emergency contraception 48 hours after an episode of unprotected intercourse. This is the third such request in a year. What problems do you face?

INTRODUCTION

This is a frequent problem in general practice. There are currently plans to make post-coital contraception available over the counter.

FIVE POINT PLAN

Issues for the patient

- Embarrassment and fears over possible pregnancy, STDs.
- Attitudes to failure of treatment.
- Explore understanding of the use of contraception.
- Underlying agenda.

Issues for the doctor

- Own personal moral code or religious objections.
- Need to sort out immediate clinical need.
- Assessment with appropriate history.
- Careful explanation and written material.
- Good record-keeping.

Risk-taking behaviour

- Explore reasons for repeated unprotected intercourse.
- Presence of multiple partners, awareness of HIV.
- History of recreational drug use.

Safety-netting

- Offer follow-up as agreed.
- Explanation of treatment e.g. failure rate, vomiting.
- Careful choice of future contraception.

Wider issues

- Deregulation of emergency contraception.
- Review availability of contraceptive advice in the practice.
- Requirement to refer to colleague if you feel unable to help.

22. A 45 year old lady whom you have treated in the past for depression confides that she is increasingly distressed by her 17 year old son coming in late and being loud and unruly. She asks if you would talk to him. What is your response?

INTRODUCTION

Patients who consult about someone else present a unique set of challenges but also deliver the opportunity for improving insight.

FIVE POINT PLAN

Issues for the mother

- Uncover her specific anxieties and worries about her son.
- Be aware of any further agenda.
- What is the social and family background?
- Is other support available?
- Increasing stress may lead to relapse.

Doctor's agenda

- Awareness of the trust patient has in you but need to clarify her expectations.
- Need to be supportive and offer positive help.
- Evaluate how much of the problem lies within each party.
- Is this an appropriate role for a doctor? Are you looking for morbidity or being a family policeman?
- Balance possibility of providing genuine help with that of exacerbating problems and inviting personal danger.
- Could disrupt the doctor–patient relationship.
- May be time-consuming but further knowledge of the family dynamics may be very useful.

Issues regarding the son

- Is he your patient and is there mutual respect?
- Is he likely to appreciate or accept advice?
- Is there an underlying problem in need of help or is this just high spirits?
- Identify clear behavioural changes e.g. use of alcohol, possible violence.

Management options

- Problem of how to make contact in least threatening manner e.g. via mother/phone/letter/home visit.
- Problem of sensitively raising the issue.
- Consider offering to see mother and son together.
- Could compromise by raising issue when he next visits surgery if appropriate.
- Explain ethical problems and duty of confidentiality to all parties.

Follow-up

- Ongoing review and support for mother.
- Other agencies such as CPN, social worker, specific support groups may be of use.

23. Discuss the points you would wish to consider when deciding whether or not to appoint a generic counsellor to your practice. Refer to the available evidence.

INTRODUCTION

There has been an astonishing recent increase in demand for counselling services in general practice despite a dearth of good evidence for its efficacy. There has been an assumption that it reduces secondary referral.

FIVE POINT PLAN

Issues for the patient

- Perceived benefit of extra service.
- May avoid need for medication.

Issues for the doctor

- Constraints on time.
- Personal need for refining consultation skills.
- Use of "active listening" while taking charge of the session.

Issues for the practice

- Appropriate selection of patients for counsellor.
- Requires available room for private use.
- Recognized accreditation of counsellor.

Resources

- Cost of hiring counsellor.
- Need to audit the benefit to the practice.
- Possibility of prescribing savings (evidence inconclusive).

Evidence issues – key points

- Lack of clear evidence of efficacy.
- Lack of evidence of cost-effectiveness.
- High rate of satisfaction from patients and GPs.
- More impressive counsellors are more verbally active rather than just listeners.
- Writing therapy (disclosure of traumas onto paper and bringing them to consultation) under care of GP alone is cost-effective.
- Recent survey shows higher referral rates to clinical psychologist after counselling and no difference in outpatient psychiatry.
- Insufficient evidence for the use of generic counsellors alone in major depression.

RELEVANT LITERATURE

Relationship between practice counselling and referral to outpatient psychiatry and clinical psychology. Cape J. *BJGP* **48:** 1477–1480, 1999.

A randomised controlled trial and economic evaluation of counselling in primary care. Harvey I *et al.* *BJGP* **48:** 1043–1048, 1998.

Should general practitioners refer patients with major depression to counsellors? A review of published evidence. Churchill R, *BJGP* **49:** 738–743, 1999.

24. List the advantages and disadvantages of GPs carrying out thrombolysis? What evidence is available to decide whether it is an appropriate activity for GPs?

INTRODUCTION

Thrombolytic therapy reduces mortality from myocardial infarction. The sooner it is received the better. The British Heart Foundation has set achievable limits of 90 minutes.

LIST

Advantages

- Speed of access.
- Proven benefit in mortality.
- Support by many hospital specialists.
- Rapid pain relief.

Disadvantages

- Special training need.
- Lack of access to defibrillator.
- Needs motivated, skilled GPs.
- Addition to already high GP workload.

EVIDENCE

For intervention

ISIS-2 showed the benefit of early thrombolysis. Streptokinase reduced death from MI by 25%. The addition of aspirin reduced it by a further 25%.

For GP role

Trials for GP thrombolysis: **GREAT** – the Grampian Region Early Anistreplase Trial of 311 patients receiving therapy a median of 139 minutes earlier than those in hospital reduced infarct size and improved left ventricular function. Mortality at 3 months was improved. When time to reach hospital exceeds 30 minutes, GP thrombolysis is appropriate.
The other large studies in this area, **EMIP** (European Myocardial Infarction Project 1993 of 5469 patients) and **MITI** (The Myocardial Infarction Triage and Intervention in Seattle) were both inconclusive for the benefit.

For GP attitudes

Most GPs in the GREAT trial were convinced by the benefits of community thrombolysis and most would be willing to perform an ECG. However, most would require further training and considered the support of the local cardiologist important.

RELEVANT LITERATURE

Call to needle times after acute myocardial infarction in urban and rural areas in north-east Scotland: a prospective observational study. Rawles J, *et al.*, *BMJ* **317:** 576–578, 1998.
Should general practitioners give thrombolytic therapy? *DTB* **32(9),** 1994.
Attitudes of general practitioners to pre-hospital thrombolysis. Rawles J. *BMJ* **309:** 379–382, 1994.

25. Outline the role of the GP in looking after stroke patients. Refer to recent evidence to support your answer.

INTRODUCTION

Cerebrovascular disease is the third most common cause of death in the developed world. The incidence is rising due to the elderly population. A GP will typically see four new strokes per year. Twenty-four per cent of the severely disabled are stroke victims.

FIVE POINT PLAN

Clinical management

- Attend promptly.
- Relevant medical examination and assessment of disability.
- Considerate liaison with family.
- Consider referral to hospital or fast-track assessment clinic.

Rehabilitation services

- Tailored, goal-directed use of coordinated in-patient and out-patient services.
- A designated unit is the gold standard of care.
- Aims agreed by patient, carers and family.
- Access to disability aids.
- Early referral for occupational therapy assessment and physiotherapy reduces disability.
- Speech and language therapy for dysphasics improves recovery.

Social issues

- Supporting the carer e.g. support groups, respite care.
- Awareness of financial aspects such as benefits.
- Advice regarding driving.
- Prospects of going back to work.

Follow-up

- Ongoing need for information e.g. stroke association.
- Consider patients' changing needs and environment.
- Medical review including use of antihypertensives/long term aspirin.
- Lifestyle review and control of risk factors e.g. smoking.
- Look for symptoms of depression.
- Post-discharge hospital review.

Related evidence

- Benefit of lifestyle changes especially smoking cessation.
- Organized stroke rehabilitation improves short and long term mortality by 25% and reduces long term disability.
- Greatest recovery occurs in the first 3 months and rehabilitation should be started as soon as the patient's condition permits.

- Home care plans have not been shown to influence admission rates or improve outcome at 6 months.
- Long term aspirin reduces further serious vascular events by 25%. It should be prescribed as early as possible in ischaemic stroke.

RELEVANT LITERATURE

Collaborative systematic review of the randomised trials of organised inpatient (stroke unit) care after stroke. *Stroke Unit Trialists' Collaboration BMJ* **314:** 1151–1159, 1997.

Diagnosis and management of stroke. *Update Mar* 1998.

26. Your practice feels it can improve the management of UTI in children. Do you feel that this is an important priority? Discuss your approach using evidence where necessary.

INTRODUCTION

UTI affects 3–5% of girls and 1–2% of boys during childhood. The consequences include chronic renal failure and long-term dialysis can be severe and the diagnosis is easily missed in the young. Guidelines exist to aid the GP.

FIVE POINT PLAN

Issues for the doctor

- Awareness of guidelines.
- Careful assessment and consideration of diagnosis.

Issues for the practice

- Agreed method and organization of urine collection.
- Ease of availability of diagnostic tests.
- Availability of a microscope within the practice.
- Preparation of written advice for parents.

Clinical management

- Need to prove diagnosis: use of microscopy, urinalysis, MSU.
- Protocol for immediate management.
- Detection of underlying cause: referral for investigation, e.g. is direct access to ultrasound DMSA scan possible?

Follow-up

- Need for prophylactic antibiotics.
- Need for post-treatment sample.
- Care shared with specialist.

Related evidence

- Royal College of Physicians guidelines: any infant with fever >38.5°C with no obvious source requires a fresh MSU sample to be taken.
- Long term antibiotic therapy does prevent further infections.
- Very low risk of renal damage if UTI after 4th birthday.
- In a 1997 study, 74% of GPs were unaware of the 1991 guidelines.
- Simple steps reduce the chance of a contaminated sample.

RELEVANT LITERATURE

The management of urinary tract infection in children. *DTB* **35(9),** 1997.
Guidelines for the management of acute UTI in childhood. *Royal College of Physicians,* 1991.
The struggle to diagnose UTI in under 2 years in general practice. Winkens RA, *et al., Family Practice* **12(3):** 290–293, 1995.

27. List some ways in which GPs can identify patients who abuse alcohol. Mention any evidence which can illustrate how effective GPs are in reversing the problem.

INTRODUCTION

Alcohol causes 28 000 excess deaths per year. Six per cent of men drink more than 50 units per week.

ANSWER PLAN

GPs can be alerted to the presence of problem drinking in several ways:

- by the presenting complaint;
- by simple opportunistic enquiry regarding lifestyles;
- by simple screening tools such as the CAGE questionnaire;
  - (i) ever felt you should **C**ut down on your drinking?
  - (ii) have people **A**nnoyed you by criticising your drinking?
  - (iii) have you ever felt **G**uilty about your drinking?
  - (iv) have you ever had an **E**ye-opener?

This can be followed up by use of a drink diary.

- by physical symptoms, any system of the body can be affected e.g. gastrointestinal effects, withdrawal symptoms, neurological symptoms;
- by psychological symptoms e.g. depression, drug abuse, behavioural changes;
- following accidents or violence;
- by third party presentation of family members;
- through awareness of domestic violence;
- awareness of other social manifestations e.g. financial or family problems;
- by examination e.g. smelling of alcohol;
- by investigation e.g. raised MCV, altered LFTs;
- other complications.

Relevant evidence

- Knowledge of guidelines: maximum recommended intake of 3–4 units per day for men, 2–3 for women (Dept of Health, Dec 1995).
- Brief intervention by the GP can reduce consumption in heavy drinkers by 20–30%; non-specialist care was as good as specialist. Reduction can be sustained at 12 months. This is cost-effective and has potentially enormous health benefits.
- Primary prevention in 8–25 year group not shown in randomized controlled trials to be particularly effective.
- GPs have shown reluctance to treat alcohol problems or attend training in the area due partly to pessimism at the outcome and other pressures of work.

RELEVANT LITERATURE

Tackling alcohol misuse: opportunities and obstacles in primary care. Deehan A, *et al.*, *BJGP* **48:** 1779–1782, 1998.

Effectiveness of general practice intervention for patients with harmful alcohol consumption. Anderson P, *BJGP* **43:** 386–389, 1993.

28. Screening for colorectal cancer is a priority in this country. Compare and contrast different methods referring to the evidence.

INTRODUCTION

Colorectal cancer is the second most common cancer fatality with 20 000 deaths per year in the UK. The overall 5-year survival is 37%.
Four methods of screening are possibilities, digital rectal examination (DRE), faecal occult blood (FOB) testing, sigmoidoscopy and colonoscopy. The search for a practical tool centres on the first three.

COMPARISON *(think of Wilson's criteria!)*

- DRE is easy but of limited clinical use
- Sigmoidoscopy

Advantages	Disadvantages
Highly sensitive and specific	Expensive
Investigation and treatment combined	Invasive
Quick	Possibly unacceptable to patients
Could reduce mortality	Operator-dependent

Related evidence

It has been shown that a single flexible sigmoidoscopy at the end of the sixth decade is a cost-effective, acceptable way of preventing 3500 deaths. Flexible sigmoidoscopy is more comfortable, is passed higher than rigid sigmoidoscopes to reach 60% of cancers.
Polypectomy after rigid sigmoidoscopy reduces the chance of rectal cancer as much as 30 years later.

- FOB

Advantages	Disadvantages
Cheap and easy	Low sensitivity
Detects blood from high in GI tract	Bleeding is a late stage
Reasonably acceptable	May miss intermittent bleeds

Related evidence

One practice in Birmingham sent Haemoccult cards to 50–80 year olds and had a 55% response finding 12 cancers. Staff were enthusiastic and the test was acceptable.
In one study, of 31 000 45–75 year olds, testing three stools reduced mortality from colorectal cancer by 17%.

RELEVANT LITERATURE

Faecal occult blood testing reduces colorectal cancer mortality (systematic review) Towler BP, *et al. EBM* **4(1),** 1999.

Screening for rectal cancer: a general practice-based study. Marjoram J, *et al., BJGP* **46:** 283–286, 1996.

29. Mr Albert is 55 years old. He comes to you requesting the blood test for his prostate. How do you proceed?

INTRODUCTION

Prostate cancer is the second commonest cancer in men. Five year survival is 43% and 30% of men have a latent cancer at death. There is no currently accepted screening programme.

FIVE POINT PLAN

Issues for the patient

- Reasons for request.
- Presence of symptoms.
- Expectations of the test.

Issues for the doctor

- Awareness of evidence.
- Good consultations skills to understand concerns of patient.
- Explanation of the test.
- Evidence-based decision on whether to offer it.

Issues for the practice

- Consider developing practice guidelines on performing blood test.
- Decide how to act on results.
- Availability of practice nurse phlebotomy.
- GP role in researching the value of testing.

Follow-up

- Review and explanation of test result if necessary.
- Discussion of ongoing concerns.

Evidence base

- No randomized controlled trial has shown that routine PSA (prostate-specific antigen) screening improves lifespan or quality of life.
- A PSA level above 6.5 ng/ml in studies has appeared appropriate for referral but the low positive predictive value has implied that the test should not be done in the absence of symptoms.
- The most cost-effective approach may be a combination of digital rectal examination and PSA but the high false positive rate is worrying.
- The level of morbidity and latency currently weigh against routine screening.
- Informed consent decreases patient interest in PSA screening.

RELEVANT LITERATURE

Detection of prostatic cancer (editorial). Schroder FH, *BMJ* **310:** 140–141, 1995.

The impact of informed consent on patient interest in PSA screening. Wolf AM, *et al., EBM* Nov/Dec: 203, 1996.

30. Discuss with reference to the evidence the treatment of depression in general practice.

INTRODUCTION

Depression is a widespread and commonly overlooked diagnosis. Primary care physicians are ideally placed to find and treat it.

FIVE POINT PLAN

Issues for the patient

- Presentation with physical complaint.
- Problems of social isolation.
- Different cultural beliefs.
- Relevant life events.

Issues for the doctor

- Better consultation skills improve likelihood of recognizing depression.
- Use of open questioning and the need for longer consultation.
- Possible underlying physical illness/multiple pathology.
- Requires high index of suspicion.
- Use of questionnaires and depression scales to aid diagnosis.

Therapeutics

- Drug versus non-drug therapy.
- All antidepressants are equally effective.
- Different side-effect profiles.
- Use of intervention by counsellors.

Practice issues

- Consider need to appoint counsellor.
- Discuss desire for a formulary.
- Extension of consulting times to improve pick-up rate.

Evidence

- About 50% of people with depression are not recognized as such by the GP.
- The 1992 Defeat Depression Campaign (RCGP consensus statement) advocated checklists for depression, education of GPs, longer consultation and use of questionnaires.
- Cognitive therapy has been shown to be as effective in the GP setting as antidepressants but time-consuming.
- Quick screening questions can be used as effective screening tool.

RELEVANT LITERATURE

Improving the treatment of depression in primary care. Moore RG, *BJGP* **47:** 587–590, 1997.

Recognition and management of depression in general practice: consensus statement. Paykel ES, *BMJ* **305:** 198–202, 1992.

31. List the advantages and disadvantages of using a prescribing formulary in general practice.

INTRODUCTION

Various formularies have existed for some time e.g. BNF, MIMS. General practice formularies have also been published.

A preferred list of medications that are agreed for use in the surgery may benefit a practice. GPs are responsible for 75% of NHS prescribing.

LIST

Advantages

- Use as an educational tool.
- Helps prevent drug-induced illness.
- Cuts prescribing costs.
- Encourages generic prescribing.
- Encourages liaison with pharmacists.
- Promotes contact with hospital consultants.
- Opportunity to audit prescribing.
- Can put level 3 PACT data into practice.
- Consistency (medication agreed by partners).
- Aid for registrars/locums.
- May be able to access via computer.
- Could help practice nurse prescribing e.g. dressings.

Disadvantages

- Rigidity – prevents adoption of new ideas.
- May limit personal armoury.
- Not always beneficial costwise.
- Hospitals may use different drugs.
- Time-consuming.
- Resistance to change and difficulty agreeing.
- Requirement for updating regularly.
- Patients may already be on different drugs.

RELEVANT LITERATURE

Constructing a practice formulary: a learning exercise. *DTB* **29(7)**, 1991.

32. A 65 year old widow has been a patient of yours for several years. On visiting her one day she presents you with a gift of a new sports car and expresses her gratitude to you. How would you react?

INTRODUCTION

Gifts are not uncommon in general practice but the size of this present would raise some controversial issues.

FIVE POINT PLAN

Issues for the patient

- Are there ulterior motives?
- Is this an attempt to get "private-style" healthcare from the NHS?
- Possibility of attempt at bribery.

Issues for the doctor

- Recognition of feelings of flattery and embarrassment.
- Worry that the gift is unsuitably large.
- Could be a presentation of a change of behaviour (assess mental state)?
- Need to be seen as incorruptible.
- Keep good records against accusations that could follow.
- Duty to act "in the patient's interests".

Issues for the practice

- Could this reflect badly on the practice?
- Inform partners/practice manager/receptionists.
- May need to consider removal from list.
- Review practice policy on gifts e.g. need to keep a log book.

Communication

- Show gratitude but careful explanation with patient of your concerns.
- Danger of disintegration of the doctor–patient relationship.
- Need to be firm and clear.
- Suggestion of other course of action e.g. donation to charity.

Medico-legal and ethical issues

- Ensure the patient is competent. (Gifts must not be accepted in the face of incompetence). Have there been other signs of disinhibited behaviour?
- Consultation with your defence union and/or GMC.
- Personal ethics: would you be happy if all your patients knew about this or if it appeared in the tabloids?

RELEVANT LITERATURE

Good Medical Practice. *General Medical Council*, 1998.

33. You are visited by a 30 year old mother and her 4 year old son who has previously been diagnosed by a specialist with a hyperactive disorder. She complains that she has just been asked to remove him from nursery due to his disruptive behaviour. She appears at the end of her tether. How do you approach the problem?

INTRODUCTION

One in 200 children develop Attention Deficit Hyperactivity Disorder. It affects males four times more frequently than females and the diagnosis requires features of inattention, overactivity and impulsiveness.

FIVE POINT PLAN

Parental factors

- Assess mother's expectation of the doctor.
- Impaired ability to cope.
- Underlying emotional/medical problems.

Consultation issues

- May be challenging and time-consuming.
- The need for good communication skills to assess where the problems lie.
- Empathetic style.
- Observe mother–child interaction.
- Simple advice to avoid triggers and encourage routine.
- May need to terminate a lengthy consultation and arrange follow-up.

Social history

- Try to assess environmental factors and level of disruption at home.
- Support network and family dynamics.
- Is housing adequate?

Primary health care team

- Attempt to get information from everyone involved in the case e.g. nursery teacher.
- Additional input from health visitor.
- Need to strongly consider specialist referral for behavioural therapy etc.

Wider issues

- Availability of resources and proper supervision of appropriate prescribing.
- Are there support groups available locally?
- Awareness of new treatments.
- The need to keep up-to-date regarding a common but difficult area.

RELEVANT LITERATURE

The management of hyperactive children. *DTB* **33(8),** 1995

Attention deficit hyperactivity disorder – a review. Williams C, *et al.*, *BJGP* **49:** 563–571, 1999.

34. A 65 year old retired railway worker with secondaries from his inoperable rectal carcinoma is under your care. He has conveyed his feelings that he is aware he is dying and wishes to do so peacefully at home and not to suffer unnecessarily. He is on the highest dose of morphine that you have encountered and is still bed-bound and in pain. You call to visit. What is your approach?

INTRODUCTION

Care of the terminally ill affects all GPs. It involves careful consultation skills, effective teamwork and knowledge of palliative care with continual reassessment as agreed with the patient and his attendees.

FIVE POINT PLAN

The doctor–patient relationship

- The need to respond to patient's personal desires and needs.
- Patient may fear loss of dignity and self control.
- Assess for signs of depression.
- Be open and supportive but clear that you cannot hasten death.
- Ongoing reassurance of relief of pain.

Treatment issues

- Symptom control is clearly inadequate and will need review.
- May require NSAID or opiate syringe driver.
- Assess need for anxiolytic or antidepressant.
- Attention to constipation, mouth care, pressure points etc.

Family issues

- Be sensitive to the input from spouse and family.
- Answer any questions and respond to their concerns.
- Awareness that duty of confidentiality to patient still exists.

Primary care teamwork

- Encourage input from all members of the multidisciplinary team.
- Named co-ordinator of care is usually the GP.
- The district nurse will be of particular importance.
- Plan for contingencies.
- Good communication regarding your plans on visiting.
- Need for coordination with specialist palliative care.
- Need for ongoing training in palliative care.
- Back-up from Macmillan nurses or Marie Curie nurses.

Legal/ethical issues

- Awareness that doctors are not able to end life at patients' request.
- Euthanasia is illegal in the UK at present although in some countries is not prosecuted if performed according to protocol.
- Consider personal ethics if asked for advice on how to take own life.

RELEVANT LITERATURE

ABC of Palliative Care: Communication with patients, families and other professionals. Faulkner A, *BMJ* **316:** 130–132, 1998.

35. One of your partners continually turns up late for his surgeries. He refuses to see extra patients at the end of surgery and you are increasingly having to cover for him. You arrange a meeting with the rest of the partners. What will you wish to discuss?

INTRODUCTION

Irregularity of a colleague's behaviour may indicate an underlying cause and could have far-reaching consequences for patient care and service delivery.

FIVE POINT PLAN

Approaching the partner

- Recognizes difficulty of combining business and personal relationship.
- Diplomatic technique, perhaps nominating least threatening doctor.
- Understand and acknowledge partner's point of view.
- Assess underlying reasons sympathetically e.g. family or work problems.
- Has he or she contacted help e.g. own GP?

Problems for the partnership

- Additional burden on other staff.
- Concerns for the future of the practice.
- The need to maintain supportive relationship between all partners.

Practice organization

- May impact provision of out of hours care.
- May require temporary leave of the partner.
- Communication strategy for staff and patients.
- The need for a locum.
- Are there underlying problems within the practice which should be re-examined?

Medico-legal and ethical issues

- Duty of confidentiality.
- Patients may have instituted complaints proceedings.
- Duty of care and the need to protect patients.

Plan of action

- Try and agree plan for the future e.g. time off, stress management.
- Reassess regularly.
- Make contingency plans to cover workload.
- Consider need to involve other agencies e.g. National Counselling Service for sick doctors, BMA's hotline for distressed doctors, GMC.

RELEVANT LITERATURE

Good Medical Practice. *General Medical Council*, 1998.

36. You have been elected to review how your practice looks after its asthmatics. Referring to evidence in your answer, outline how you will proceed.

INTRODUCTION

The burden of asthma morbidity is increasing despite falling mortality rates in the UK and there is evidence of poor asthma management by both doctors and patients. Nevertheless, most patients with long-term asthma can be managed in general practice.

FIVE POINT PLAN

Define aims

- Discussion with partners to agree aims and enlist cooperation as necessary.
- Review current guidelines.

Define methods

- Opportunistic approach or identification of asthmatics using records.
- Describe methods of patient education.
- Review of notes for assessment of disease control.
- Select parameters of further assessment e.g. the relevant PACT data or adequate inhaler technique etc.

Practice organization

- Consider use of nominated practice nurse to interview patients.
- Decide on whether to put specific time aside.
- Is appropriate equipment available e.g. for testing inhaler technique?
- Ensure patients are aware if a new service is started.

Evidence-based approach and the use of guidelines

Much has been written but several issues of evidence are applicable here

- Application of the BTS guidelines suggest a stepwise approach to care, the use of written self-management plans with audio-visual reinforcement and consideration of prescribing a peak flow meter.
- The BTS has also produced guidelines on the use of home nebulizers particularly for acute severe asthma and the elderly.
- Holding chambers can be as effective as nebulizers.
- Limited educational interventions increase knowledge of asthma but not health outcomes.
- Use of written plans is more effective.
- Over-reading from peak flow meters may cause undertreatment.
- The Royal College of Physicians now recommends three simple questions at every asthma consultation to assess uncontrolled asthma. This is a good audit tool and responses should be recorded on computer if possible.
  (i) have you had difficulty sleeping due to asthma symptoms?

(ii) have you had your usual asthma symptoms during the day?
(iii) has asthma interfered with your usual activities?

Follow-up and audit

- Meet with partners to agree changes.
- Audit data on changes in prescription patterns/cost, admission rates or retention of advice.
- Consider development of protocol.
- Develop patient leaflet/use of written instructions.

RELEVANT LITERATURE

British Thoracic Society: The British guidelines on asthma management. *Thorax* **52:** (Suppl. 1) S1–21, 199 7.

Long term management of asthma in adults. *Prescriber's Journal* **39(3),** 1999.

Self management education for adults with asthma improves health outcomes (Cochrane Database meta-analysis), *EBM* **4(1),** 1999.

Guided self-management of asthma – how to do it. Lahdensuo A. *BMJ* **319:** 759–769, 1999.

37. What do you understand by the term "guidelines"? List the benefits of using them and outline how you would construct a set of guidelines for your practice.

INTRODUCTION

Guidelines are defined as systematically developed statements to assist practitioner and patient decisions about appropriate heath care for specific clinical circumstances. The National Institute of Clinical Excellence is taking a key role in the production of authoritative guidance.

LIST OF BENEFITS

- Explicitly state practical targets.
- Ensure consistency in the primary care team.
- Can turn knowledge into action by making best and clearest use of data.
- A useful aide-memoire.
- A useful guide for the newcomer to the practice e.g. registrar.
- Can cover all aspects of care e.g. investigation, diagnosis, treatment.
- Evidence has shown in several studies that they can improve delivery of care.
- Likely to be used by those who have helped develop them.

CONSTRUCTION OF GUIDELINES

Decide on objectives

- Identify a clear need.
- Get agreement from those who will use them.
- Decide what needs to be achieved.

Select approach

- Do adaptable guidelines already exist? (Many web sites have useful guidelines already available.)
- Best available evidence base is the most widely accepted and most valid basis for guidelines.
- Relevant high quality sources are researched, reviewed, critically appraised and summarized.
- Relevant expertise e.g. hospital consultants, librarians, health economists should be consulted as necessary.
- Other less rigorous (and less valid) approaches may take the form of small surveys or consensus opinion.
- One approach may involve central development with regional adaptation to allow "local ownership" of the guidelines.

Writing the guidelines

- The gold standard should incorporate a full account of the risks and benefit of treatment, patient preference and cost-effectiveness.
- Should be clear and appropriate to need.

- Statement of aim and method.
- Statement of who is responsible for each component of care.
- Describe monitoring and recording.
- Describe standards and targets.

Dissemination

- Guidelines are useless unless distributed effectively.
- Period of discussion, debate and adaptation.

Audit and update

- Assess for improvement in delivery of patient care.
- A date for review should be included as evidence may change.

RELEVANT LITERATURE

Construction and use of guidelines. *Prescriber's Journal* **39(3),** 1999.
General practitioner's use of guidelines in the consultation and their attitudes to them. Watkins C, *et al., BJGP* **49:** 11–15, 1999.

38. Ben is 4 years old and has had no success with toilet training. Instead of using the potty, he regularly soils his clothes and the house and conflict is arising in the household. His mother attends and insists it has to stop. What are the important areas of this consultation?

INTRODUCTION

Soiling can be a symptom of turmoil in the child and a cause of turmoil in the household. A sensitive and considered approach is needed to reach the likely cause.

FIVE POINT PLAN

Parental feelings

- Explore mother's concerns and expectations.
- How is she coping?
- Possibility of a hidden agenda.

The consultation

- Empathy regarding the unpleasantness of the situation.
- Recognition of mother's anger.
- Sympathetic communication skills to ease tension.
- Explore habits, routines, diet and timing of the problem.
- Examine old notes for other clues e.g. regular attendances.
- Relevant clinical examination and assessment of development.
- Need to manage own time efficiently.

Family dynamics

- Are there any new stresses in the household?
- Evaluate role of father.
- Assess social circumstances and what support is available for the mother.

Primary Health Care Team

- Close relationship with local health visitor is required.
- Discussion with nursery teachers and/or community child health.

Clinical management

- Develop differential diagnosis (e.g. constipation with overflow, behavioural response, possibility of child abuse).
- Start treatment plan e.g. laxative if suspicious of constipation and follow-up.
- Consider star chart for using the toilet.
- Consider referral for further management/behavioural approach as necessary.

39. List some of the evidence that helps the GP in the treatment of acute low back pain.

INTRODUCTION

The overall cost of low back pain in 1993 was estimated at nearly £6 billion. Apparent disability from back pain has increased fourfold since the late 1970s. With over 14 million GP consultations a year, it clearly merits a carefully considered approach.

LIST

Clinical assessment

- Diagnostic triage forms the basis of further investigation and referral to detect uncommon underlying sinister pathology.
- Psychosocial factors influence disability and response to treatment: eligibility for benefits is a predictor of persisting pain

Advice

- Early return to normal physical activity leads to the more rapid recovery.
- Minimize bedrest. Use for pain relief alone.
- Muscles that are not used lose 3% of their strength per day

Medication

- Use analgesia at regular intervals.
- No clear evidence that co-proxamol or co-dydramol are superior to paracetamol alone.
- Short term use of muscle relaxants are more effective than placebo

Interventions

- Back exercises are no more effective than no treatment initially but may benefit those who have not returned to work at 6 weeks.
- Some may benefit from manipulation or mobilization by a physiotherapist or chiropracter.
- Acupuncture can be of use but is not widely available on the NHS.
- General exercise reduces illness behaviour.

Guidelines

- The RCGP guidelines outline the criteria to detect serious spinal pathology e.g. age >55 years, gradual onset, concurrent malaise, cauda equina symptoms etc.
- Detailed reassessment needed after 3 months including physical, psychological and social aspects.
- Royal College of Radiologists' guidelines suggests X-ray is not indicated in the absence of sinister pointers.

RELEVANT LITERATURE

Managing acute low back pain. *DTB* **36(12)**, 1998.
Clinical guidelines for the management of acute low back pain. *RCGP* 1996.
Systematic reviews of bed rest and advice to stay active for acute low back pain. Waddell G, *et al., BJGP* **47:** 647–652, 1997.

40. A 20 year old man asks you for an "AIDS test". Describe the important aspects of your reply.

INTRODUCTION

The GP is ideally placed for pre-test HIV discussion. Recent guidelines have helped to demystify the procedure.

FIVE POINT PLAN

Patient's reasons

- What are the reasons for presenting now?
- What are the main concerns?
- Check understanding that it is a test for HIV rather than AIDS.
- Is there a hidden agenda?

Target for the consultation

- Identify risk activity.
- Consider the patient's sexual relationships.
- Take a related history including travel, drug use, occupation.
- Discuss pros and cons of testing.
- Discuss the impact of a positive test.
- Is there an underlying psychological problem?
- Explain the procedure.
- Suitable recording of salient aspects.
- Arrange appointment to discuss results.

Consent

- Obtain informed consent.
- Ask the patient to summarize or repeat salient points.
- Ask if they want to go ahead.

Confidentiality issues

- Consider coded labelling of samples.
- Avoid giving results over the phone.
- In the event of a positive test, you may need to explain the importance of informing your fellow professionals to avoid a breach of confidentiality, e.g. for a future surgical referral.

Issues for the practice

- Is written material available?
- Ensure failsafe follow-up of positive results.
- The need to develop a protocol.

RELEVANT LITERATURE

Guidelines for pre-test discussion on HIV testing. *Dept of Health,* 1996.
Discussing HIV testing in general practice. *Pulse* Nov. 1996.
Serious communicable diseases. *General Medical Council* 1997.

41. Summarize the evidence that allows a rational approach to diagnosis and management of sore throat in general practice.

INTRODUCTION

In most cases the condition is minor and self limiting and may not present to the GP but a significant number get unacceptable morbidity, loss of earnings etc. Antibiotics of questionable efficacy are therefore often handed out by GPs.

EVIDENCE

Presentation

Clinical examination is unreliable for differentiating between viral and bacterial infection. It lacks sufficient sensitivity and specificity.
Studies on the predictive value of constellations of symptoms are conflicting and inconclusive.
Underlying psychosocial influences should be considered.
Studies show that GPs often perceive more pressure to prescribe than actually exists.

Diagnosis

Common testing methods include throat swab and rapid antigen testing. Anti-streptolysin O titre is a gold standard for research but with no practical clinical value.

Throat swabs

- Should not be carried out routinely in sore throat.
- Are not cost-effective (costs £4 and report takes 24–48 hours).
- Detect carriage as well as infection.
- Negative culture does not rule out streptococcal infection.
- Can medicalize illness.

Rapid Antigen Testing (near patient testing)

- Should not be carried out routinely in sore throat.
- Costs £4 and takes 10 minutes to report.
- Sensitivity varies between 61% and 95%.
- Use of the test changes prescribing patterns very little.

Treatment

Use evidence-based guidelines if possible e.g. those produced by SIGN which suggest the following:
Admission: Serious signs such as stridor merit admission to hospital.
Analgesics: Paracetamol is the agent of choice for all ages.
Antibiotics: Evidence does *not* support routine use to:

- Reduce symptoms.
- Prevent rheumatic fever or glomerulonephritis.
- Prevent cross infection in the general community.
- Prevent suppurative complications.

Any marginal benefit of antibiotics tends not to be translated into cost-effectiveness. Side effects include anaphylaxis and death, candidiasis, diarrhoea and unwanted pregnancy.
Ampicillin-based antibiotics should be avoided first line due to chance of infectious mononucleosis.
The preferred regime when treating (suspected) Group A β-haemolytic streptococcus is 10 day course of penicillin q.d.s. which may reduce recurrence.
Referral for tonsillectomy is indicated when episodes are disabling, more than five per year and symptoms have been continual for at least a year.
Education: Patient information leaflets may empower home management in future.
Prescribing has been shown to increase reattendance.

RELEVANT LITERATURE

Management of sore throat and indications for tonsillectomy. *Scottish Intercollegiate Guidelines Network* http://www.show.scot.nhs.uk/sign/home.htm) Jan. 1999 (an excellent review).
Diagnosis and treatment of streptococcal sore throat. *DTB* **33(2),** 1995.

42. Anna is a 14 year old girl who attends with her 25 year old boyfriend. He asks you to prescribe the oral contraceptive pill as they are planning on sleeping together. The family are personal friends of yours. What issues are raised?

INTRODUCTION

The UK has the highest teenage pregnancy rate in Western Europe. At present 19% of women start sexual intercourse before the age of 16.

FIVE POINT PLAN

Problems for the patient

- Embarrassment about presenting.
- Is she feeling pressured by her older boyfriend?
- Explore patient's knowledge and understanding.

Issues for the doctor

- Potential conflict of interests as you know the family.
- Concern over the large age gap and greater experience of boyfriend.
- The need to talk to Anna alone.
- Encourage her to talk to parents and perhaps return to see you with them.
- Offer to talk to them on her behalf.
- The need to be able to justify your actions before the GMC.

Use of the consultation

- There is a lot to get through and they may default at follow-up.
- Sympathetic and non-judgmental manner.
- Take the opportunity to encourage a sense of self-confidence and personal worth.
- Appropriate medical history and examination as for any such prescription.
- Explain risk of HIV, STD and relationship of early intercourse and cervical dysplasia.
- Emphasize that she informs her parents and explore why she may be reticent.

Legal and confidentiality issues

- What is proposed is technically rape even if she consents.
- Has abuse already taken place? The right of confidentiality must be balanced against the protection of vulnerable people from serious harm.
- Warn the boyfriend he is about to break the law.
- Reassure Anna of your duty of confidentiality owed to the patient.
- Requirement to ensure her understanding and intelligence is sufficient to allow consent.
- Patient's concern about confidentiality is the main deterrent to asking for contraception. Nearly 75% of the under 16 year olds fear GPs would not guarantee confidentiality.

Management plan

- Act in the patient's best interests.
- Justify your decision on the prescription being aware that intercourse is likely still to occur.
- If you feel unable to agree you must refer to a colleague or advisory service (e.g. Brook Youth Advisory Centres) with different views.
- Consider the need for further counselling about the psychological and physical implications.
- Offer follow-up.

43. General practice is a stressful job. List the stresses you may encounter and outline how you may deal with them to maintain a high standard of patient care.

INTRODUCTION

Evidence has shown a high satisfaction for the type of work of general practice but this drops sharply when stress is perceived in personal and work-related areas. It is high in young idealists and patient-centred GPs.

LIST OF STRESSES

- Increased workload.
- Constant interruptions at work.
- High patient expectations.
- New contracts and regulations.
- Reduced autonomy.
- Feeling ill-prepared.
- Fear of complaints and litigation.
- Interference with family life.

MANAGEMENT OF STRESSES

- Recognition of low performance.
- Prioritize and delegate where possible.
- Aim for realistic consultation rate.
- Accept the unchangeable.
- Organizational changes e.g. fair and balanced decision-making in the practice and adjustment of work patterns.
- Develop interests in new areas to face fresh challenges.
- Open discussion of concerns.
- Time management e.g. use of protected time. Part-time GPs have a lower rate of burnout.
- Be assertive.
- Avoid self medication.
- Pursue physical activity.
- Have regular breaks.
- Plan ahead.
- Seek confidential third-party help.
    (i) Colleagues.
    (ii) Stress-management programmes.
    (iii) Own GP to allow yourself the best available care.
    (iv) Advice from LMC adviser, mentor or National Counselling Service for sick doctors.

RELEVANT LITERATURE

Avoiding burnout in general practice. Chambers R, *BJGP* **43:** 442–443, 1993
Avoidable pressures could relieve doctor's stress. Dillner L, *BMJ* **304:** 1587, 1992.

44. Tom Pope, a 50 year old patient of yours, has recently had a chest radiograph which shows inoperable neoplastic metastases. The appointment has been arranged to discuss the result. How would you handle the situation?

INTRODUCTION

Breaking bad news is an unpleasant job for all GPs but evidence has shown that clear communication skills can make an enormous long term impact on patient satisfaction. It has also been shown that the vast majority want to know their diagnosis and have a say in who else gets told.

ANSWER PLAN

Several plans are possible. You may wish to modify your favourite consultation model, for example, Neighbour's 5-point plan of connect, summarize, hand over, safety-net and housekeeping. However the following points should be included.

Preparation

- Know all the facts in advance.
- Find out who the patient wants to be present.
- Ensure privacy and comfort.

Achieve understanding

- Establish patient's knowledge and views of diagnosis – "how did it all start?"
- Give a warning shot "I'm afraid it looks rather serious".
- Speak clearly avoiding jargon.

Pacing and shared control

- Is more information wanted? (If necessary, ask the patient.)
- Allow pauses, silences and denial.
- Involve the patient in choosing pace and making decisions.

Respond to emotions

- Encourage ventilation of feelings and allow to cry.
- Use touch if appropriate. Conveying empathy is key to a satisfactory consultation.
- Be aware of verbal and non-verbal communication.

Respond to concerns and questions

- Narrow the information gap. Bear in mind detail will not be remembered.
- Identify his main concerns at the moment.
- Allow time and space for expression.

Summary and plan

- Summarize the ground covered.
- Have a plan for future treatment.
- Foster hope.

Closure

- Finish with positive points.
- Offer your availability for further clarification perhaps with family.
- Ensure patient has transport home.

45. List the advantages and disadvantages of telephone consultations. How would you structure such a consultation to avoid any problems?

INTRODUCTION

Use of the telephone in general practice is an increasingly important aspect of care. Each must be seen as a true consultation rather than merely simple advice. The term "triage" refers to a health worker (traditionally a doctor but increasingly a nurse e.g. in NHS Direct) deciding, on the basis of information collected, how best to deal with a problem.

ANSWER PLAN

Advantages

- Vast majority of diagnoses are made on history alone.
- Convenient for patients.
- Rapid access for patients if well organized.
- Can be delegated to a suitably trained nurse and free the doctor for more complicated cases or other practice duties.
- Protocols are available to guide decisions.
- Evidence has shown that triage by telephone facilitates:
  (i) Speedier access to medical services.
  (ii) Fewer home visits.
  (iii) Fewer extras in the surgery.
  (iv) Reduction in stress.

Disadvantages

- An essentially new consultation skill which requires further training.
- Loss of visual and behavioural clues.
- Sub-optimal communication can lead to misunderstanding.
- Easy to bring pre-conceptions to the consultation.
- Problems with nurse-led triage:
  (i) Time required to develop safe guidelines.
  (ii) Need for quality control and audit.

The consultation

- Introduce self.
- Record details of patient, time etc.
- Gathering information.
  (i) Gain patients' confidence.
  (ii) Systematic but positive, sympathetic approach.
  (iii) Careful relevant past history.
  (iv) Need to be in a position to make a reasonable clinical judgment.
  (v) Identify specific concerns.
- Consider likely diagnosis.
- Giving advice.
  (i) Home management plan if appropriate.
  (ii) Assess caller's satisfaction with approach.

- Safety-netting with explicit plan of when to call back.
- Record-keeping.
  (i) Record history, advice and safety-net details.
  (ii) Needs to be contemporary with the consultation.

RELEVANT LITERATURE

Safety and effectiveness of nurse telephone consultation in out of hours primary care: a randomised controlled trial. Lattimer V, *et al.*, *BMJ* **317:** 1054–1059, 1998.

Telephone triage of acute illness by a practice nurse in general practice: outcomes of care. Gallagher M, *et al.*, *BJGP* **48:** 1141–1145, 1998.

46. You call a practice meeting to suggest increasing your consultation time from 5 to 10 minute appointment slots. Your partners seem initially unhappy with this. Outline the pros and cons with reference to the evidence.

INTRODUCTION

The traditional 5 minute time slot is disappearing. The increasing demands of delivering primary care necessitate an increased consultation time.

ANSWER PLAN

Advantages

- To realize the exceptional potential of the consultation.
- Increased patient satisfaction.
- Time to find the psychological underpinnings of the consultation.
- Time for opportunistic health care.
- Requirement of proper computer use.
- Better note-taking and housekeeping.
- May lower rates of reattendance.

Disadvantages

- Extra workload may initially require increased number of surgeries.
- Five minute appointments are more available.
- Implications for re-organization of the practice e.g. extra delegation.
- Patients may consult no less frequently.

Evidence suggests

- The demand for consultations is still increasing.
- Nearly a fifth of patients are dissatisfied with the time allowed to them.
- More problems are identified in longer consultations.
- But surgeries are more likely to be overbooked.
- Booking interval has not been shown to influence the time spent.
- Some research shows that longer consultations leads to fewer referrals and less prescribing.
- GPs on 5 minute appointment consistently run late.
- And have significantly more stress!

RELEVANT LITERATURE

Consultation length in general practice: a review. Wilson A. *BJGP* **41:** 119–122, 1991.

47. The risk of violence is ongoing in general practice. What can be done to survive its impact?

INTRODUCTION

Violence is a behaviour which causes anxiety, fear and disruption in practice staff. Over 90% of GPs can expect to be verbally abused, 18% attacked and 11% injured in the line of duty. More than 97% of incidents can be resolved with good communication skills.
Reports suggest the problem is increasing.

FIVE POINT PLAN

Prevention

- Avoid prolonged waiting times.
- Ensure staff are courteous and helpful.
- Wariness of strangers e.g. new temporary resident.
- Avoid dangerous times e.g. one staff member locking up alone.
- Inform others where you are going.
- Be aware of exits.
- Take extra care on night visits.

Recognition

- Best predictor is a history of violent behaviour.
- Other predictors are severe stress, loss of social support, housing etc.
- Index of suspicion for use of drugs and alcohol.
- Observe body language: glaring eyes, loud voice, pacing, argumentative behaviour, repetitive movements, swearing, hands above the waist.

Practice initiatives

- Staff training to recognize warning signs.
- Share information on dangerous patients. Identify and tag notes.
- Prepare guidelines for contingency planning (zero tolerance stand).
- Local health authority may have guidance documents.
- Set up reporting system for untoward incidents.

Premises

- Calm, comfortable waiting area easily observed by receptionists.
- Reduce potential weapons in the consulting room.
- Ensure quick getaway.
- Designated parking near the surgery.
- Police will advise on security issues.

Defusing the situation

- Keep calm and try not to appear intimidated.
- Be open but avoid direct confrontation.
- Active listening with good non-threatening eye contact.

- Lower your voice to the extent to which the patient raises his.
- If in danger, leave the room.
- If in danger, use personal alarms/panic button/call the police.
- Legal entitlement to use reasonable force as last resort if you feel proficient.
- Always debrief and support afterwards allowing opportunity for discussion.
- Follow up the patient and explore the reasons for the incident. Some practices draw up a contract of acceptable behaviour.

RELEVANT LITERATURE

Tackling violence. Shepherd JP, *BMJ* **316:** 879, 1998.
Dealing with violence. *Pulse Mar.* 1998.

48. List the benefits that an expanded role for the pharmacist can have on the delivery of primary care.

INTRODUCTION

As part of the developments in primary care, closer bonds with other health professionals are being established.

THREE POINT PLAN

Role of pharmacists in the community

- Unique access to healthy people helps target health promotion.
- Often the first port of call for patients.
- Education of patients about medication.
- Facilitate responsible self-medication.
- Prepare monitored dosage regimens.
- Maximize health gain.
- Can provide support to patients in their homes.
- Provide collection and delivery services.
- Can provide a "medicines assessment service".
- Health promotion activities in schools etc.

Role in the practice

- Could reduce the frequency of GP consultation.
- Promote efficient and cost-effective prescribing.
- Benefits can offset employment costs.
- Rationalize repeat prescription system.
- Create a practice formulary.
- Already have experience of working towards defined business targets.
- Potential to improve patient compliance with treatment.
- Address patient concerns not voiced to the doctor.
- Relay information about patient's view of their treatment e.g. drugs returned to the pharmacy.
- Improved medicine management may reduce admission rates.

Role in primary care groups

- Have knowledge and skills to allow many varied roles.
- Helping primary care groups to achieve objectives.
- Local Pharmaceutical Committees (LPCs) are organizing a joint approach to PCGs.

RELEVANT LITERATURE

Controlled trial of pharmacist intervention in general practice: the effect on prescribing costs. Rodgers S, *et al., BJGP* **49:** 717–720, 1999.

Ten points on the evolving role of pharmacists *GP,* Oct. 1998.

49. Mr. Dan Lutman is a 70 year old patient of yours who has been suffering with Alzheimer's disease for the last few years. His wife has been coping admirably with his deterioration. You visit to review his care. How do you proceed?

INTRODUCTION

Dementia affects 5% of people over 65 years. Half of these cases are due to Alzheimer's disease. Informal carers fulfil a huge burden of care and play a crucial part in how society cares for our elderly.

FIVE POINT PLAN

Managing the patient

- Treat any concurrent illness.
- Assess degree of problems e.g. aggression, wandering, insomnia.
- Treat risk factors.
- Assess care needs.
- Consider antidementia drugs.

Supporting the carers

- Assess coping and offer support.
- Assess need for a break.
- Look for signs of depression.
- Open discussion of prognosis.
- Assess need for advice and help.
- Is appropriate help available?

Managing the environment

- Close liaison with social services and occupational therapy assessment.
- Consider safety at home e.g. presence of obstacles.
- Consider respite care.

Legal/practical issues

- Suggest the making of a will and enduring power of attorney.
- Impaired driving ability requires notification to the DVLA.

Plan for the future

- Consider the possibility of long term residential care and encourage carer to think early about this.
- Consider referral for psychogeriatric input.
- Further advice from the Alzheimer's Disease Society or local support groups.

50. Michael is a 25 year old patient of yours who was assaulted two weeks ago. He has not been himself since the attack and he attends with his wife who has insisted he comes to see you. What issues would you want to look at?

INTRODUCTION

Men between 16 and 29 years are most at risk for all types of violent crime except rape and domestic violence. Many complain of lack of information and understanding.

FIVE POINT PLAN

Effects on the patient

- Stages of reaction to the attack: shock and denial, fear, apathy and anger with guilt and depression, resolution or repression.
- Physical effects: insomnia, lethargy, headaches, decreased libido.
- Psychological effects: distress, depression, nightmares, loss of confidence, irritability, impaired concentration.
- Behavioural effects: increased smoking and alcohol consumption, social withdrawal.

Doctor's agenda

- Non-judgmental approach with appropriate reassurance.
- Detailed history e.g. of circumstances/substance abuse.
- Look for evidence of psychological reaction.
- Appropriate documentation of injuries and comprehensive record-keeping.
- Share information about options for treatment.

Effect on relationships

- The spouse and other relatives may suffer severe emotional distress also compounding the effect on the relationship.

Management

- Medical need for treatment e.g. physiotherapy if disability.
- Consider trauma counselling, brief psychotherapy or medication.
- Follow up for potential long-term effects of anxiety or depression, substance abuse. One study showed that levels of anxiety and depression after violent crime had not eased after 3 months. This emphasizes the need for prolonged support.
- Advise compensation may be available through the Criminal Injuries Compensation Authority.

Other sources of help

- Initial reporting of violence to the police may involve police surgeon.
- Contact Victim Support for free confidential emotional and practical assistance.
- Written information e.g. practice leaflet.

RELEVANT LITERATURE

Treating Victims of Crime – guidelines for health professionals. *Victim Support* 1995

51. What do you understand by the term "rationing" with regard to health care? Discuss the reasons for it and the ways in which it can occur within the NHS.

INTRODUCTION

Rationing is the inevitable reality of prioritization that bridges the gap between the needs of patients and the capacity of the health service and society to meet those needs.
General practice in its role as gatekeeper is becoming increasingly active in rationing decisions but responsibility for it is becoming more diffuse.

FIVE POINT PLAN

Three levels of intervention

- *Central level:* by the government through funding, regulation of standards and clinical governance. NICE needs to provide fully costed guidelines at a national level.
- *Local level*: by a primary care group or practice through resource allocation and cash-limitation.
- *Individual level*: by clinical freedom in the consultation.

Demand-led pressures

- Wants and needs fuelled by increasingly high expectations from e.g. The Patient's Charter, The Health of the Nation, Our Healthier Nation reports.
- Increased consumerism and openness of complaints policies.
- Therapeutic developments increase expectation.

Resource-led limitations

- Decisions can be made by management or staff.
- Availability of physicians e.g. an understaffed practice reduces delivery of patient care by having too few appointments available.
- Deterrence with prescription charges.
- Part-funding of treatments.
- Need for lifelong learning to have up-to-date primary care available.

Information-led limitations

- Information and awareness is required before a service can be used.
- Evidence-based guidelines or clinical distillations are required.
- Patients are increasingly widely informed e.g. by the Internet, to challenge the doctor's rationing.

Decision and provision

- Delivery of care is influenced by progression from reactive to proactive care and from local to national responsibility.
- Defining a limited range of services e.g. removal of NHS cosmetic surgery, dentistry, infertility treatment.
- Use of cash-limited budgets.

- Clinical governance and outcome measures will be used to explore and correct variations in care.
- New guidelines can help to ration by tackling different areas e.g. education, increasing funding and resources or managing demand.
- Difficulties arise when so many GP decisions are made in a climate of ambiguity. Clear acceptable guidelines may be elusive.

RELEVANT LITERATURE

Rationing in the NHS. *RCGP Discussion Paper* 1999.

52. What are the problems with delivering preventative health care to men? How might you reverse these in your own practice?

INTRODUCTION

Men are a difficult proposition. They attend their GP half as often as women and present late with more advanced illness. Palatable health initiatives are needed to bridge the gap.

FIVE POINT PLAN

Problems for the patient

- Men die on average 5–6 years earlier than women.
- Heart disease is the greatest single cause of early male death.
- Men die more often from rectal cancer than women.
- Men are less motivated and more resistant to health advice.
- Alcohol abuse is far higher than in women.
- No widely recognized screening programme is yet in place for men.

What to target

- Consider different areas – exercise, smoking, alcohol, cholesterol, tetanus status, BP.
- Consider pros and cons of testicular self examination, F.O.B.s in the over 50s, PSA test in men with outflow obstruction.

Initiating a well-man programme

- Nominate team members e.g. doctor, practice nurse, health visitor.
- Discuss and define objectives and develop protocols.
- Select approach (opportunistic or clinic based).
- Select methods and interventions e.g. CAGE questionnaire, brief anti-smoking advice, relaxation etc.

Practice issues

- Appropriate training courses for staff.
- Is any financial incentive available?
- Cost of administration e.g. letters of invitation, publicity.
- Provides useful basis to incorporate further guidelines.
- Useful practice registers in place for the future.

Review and audit

- Assess success and popularity of approach.
- Incorporate changes in national advice.
- Expand with use of future developed guidelines e.g. from the National Screening Programme.

53. What are the requirements of a good screening programme? Which of these apply to a new programme that screens women for chlamydial infection? Use evidence in your answer.

INTRODUCTION

Chlamydia has been recognized by the Chief Medical Officer's Expert Advisory Group as a condition which requires a detailed preventative strategy.

GENERAL REQUIREMENTS OF SCREENING

These are described by Wilson's criteria (modified to the acronym **IATROGENIC).**

- Important condition.
- Acceptable treatment for the disease.
- Treatment (and diagnostic) facilities available.
- Recognizable latent or early symptomatic stage.
- Opinions on who to treat are agreed.
- Guaranteed safety and reliability of test.
- Examination acceptable to patient.
- Natural history of disease is known.
- Inexpensive and simple test.
- Continuous rolling programme not just a one-off.

APPLICATION TO CHLAMYDIA

- It is the most common curable sexually transmitted infection in the UK with a prevalence of 3–4%. Its complications are very difficult to treat – pelvic inflammatory disease, tubal infertility and ectopic pregnancy.
- The latent stage is frequently asymptomatic but recognizable with testing.
- Screening asymptomatic women in America has led to a 56% reduction in the incidence of PID.
- Testing has become more acceptable with the development of tests based on nucleic acid amplification such as the ligase chain reaction which can be performed on urine samples avoiding the need for endocervical swabs.
- Cost-effectiveness in primary care has yet to be clearly established but seems likely to be favourable.
- Guidelines on who to screen have been suggested – all symptomatic women, those seeking TOP, women under 25 opportunistically and those over 25 with a recent change in partner. The latter has been shown to improve detection.
- Evidence of a clear benefit and of the best approach to take are awaited.

RELEVANT LITERATURE

Principles and practice of screening for disease. Wilson JMJ and Junger G. WHO Public Health Paper, Geneva 1968.

Screening for genital chlamydial infection: the agenda for general practice. (Editorial) Stokes T, *et al.*, *BJGP* **49:** 427–428, 1999.

Screening for chlamydia trachomatis. Boag F, *BMJ* **316:** 1474, 1998.

54. Mr Collins is a 58 year old patient of yours who is being cared for at home by his wife of 30 years. He was diagnosed with a brain tumour 18 months ago. He is comfortable and his care package has been carefully optimized but it becomes obvious to you that he has no more than 2 weeks to live. What help can you offer in the coming months?

INTRODUCTION

A GP with a list of 2000 patients will encounter 25 deaths per year. Between the ages of 45 and 59, 8% of women will be bereaved of a spouse.

FIVE POINT PLAN

Breaking bad news

- Careful explanation of your assessment should involve Mrs Collins' perceptions and expectations.
- Use time to achieve an understanding; communicate slowly with considerate honesty.
- Respond to her emotions and offer support.
- If possible offer a contact number to call you at home.
- Try to encourage planning for the time of death.

At death

- Attend promptly to confirm death.
- Offer practical information e.g. undertakers, registration of death, death certificate.
- Assess reaction and needs.
- Notify other caring agencies e.g. hospital consultant.
- Note in relatives' records for reference.

Initial support

- Maintain awareness that there is significant mortality in the survivor, greatest in the first 6 months.
- Praise the support and nursing care given.
- Share feelings, listen rather than talk.
- Reinforce the normality of the grief reaction (shock and denial, searching and yearning, depression and disorientation and reorganization).
- Consider anxiolytic medication.

Medium term support

- Follow-up visit to answer any questions – this can be a period of searching.
- Feeling angry and depressed is normal.
- Monitor return to normal pattern of living.
- Allow ventilation of feelings and encourage further discussion.
- Monitor for abnormal bereavement reaction.

Follow-up

- Assess need for further contact.
- After a few months, grief should be starting to resolve.
- Look for clinical depression.
- Consider use of self-help group or other referral.
- Be aware of anniversaries.

RELEVANT LITERATURE

ABC of Palliative Care: Bereavement. Sheldon F, *BMJ* **316:** 456–458, 1998.
Bereavement in adult life. Parkes CM, *BMJ* **316:** 856–859, 1998.

55. Mr Rutter is 55 years old and has come to see you after his third blood pressure measurement. The readings have averaged at 165/105. He tells you that he is not prepared to take medication. How do you proceed?

INTRODUCTION

Hypertension is an important risk factor for myocardial infarction and vascular disease but lack of compliance with medication is a widespread problem.

FIVE POINT PLAN

Patient's agenda

- Explore ideas, concerns and reasons for non-compliance.
- Denial of illness.
- May be unaware of importance.

Doctor's agenda

- Take appropriate history.
- Explore risk factors (smoking, family history, exercise, diabetes etc).
- Examination for end-organ damage e.g. fundoscopy, carotid bruit, peripheral pulses.
- Investigation e.g. urinalysis, U&E, blood sugar, cholesterol.
- Issues of keeping up-to-date with latest recommendations.

Sharing information

- Explanation of pros and cons of treatment.
- Explanation of possible side effects of medication.
- Lifestyle advice/non-drug treatment.
- Efforts directed to overcome possible non-compliance.

Practice organization

- Implementing BHS guidelines (taking BP in adults 5-yearly up to age 80 years and annual retesting in borderline values).
- Development of protocol to identify and treat hypertensives.
- Cost and organizational implications of formal screening.
- Ongoing audit of current practice.

Management plan

- Achieve common understanding.
- Agree on treatment plan.
- Use evidence-based guidelines: use low-dose thiazide or β-blockers first line unless contraindicated.
- If no agreement on medication, make substitute plan to monitor and review while emphasizing non-drug treatment.
- Arrange follow-up – planned or opportunistic.

RELEVANT LITERATURE

British Hypertension Society guidelines for hypertension management: summary. *BMJ* **319:** 630–635, 1999.

56. What points should a practice consider in changing out of hours care from a practice-based rota to the use of a deputizing service?

INTRODUCTION

A more flexible approach to the delivery of out of hours care has been essential in the face of relentlessly increasing demands and government initiatives. Job satisfaction has increased and patient care does not seem to have suffered significantly.

FIVE POINT PLAN

Changes for the patient

- Loss of continuity of care.
- Improved care from less exhausted doctors.
- Change in response times.
- Altered prescribing levels and admission rates?

Issues for the doctors

- Cost of DDS and loss of income.
- Benefits in lifestyle and reduction of stress.
- Knowledge of the prescribing, visiting, phone advice habits of the service.
- Careful consideration of alternative strategies e.g. cooperatives.

Organizational issues

- Need to monitor care provided by the deputizing service.
- Methods of measuring patient satisfaction.
- Efficient relay of information after consults.
- Need to inform changes to patients.

Legal issues

- Ongoing obligation for 24 hour responsibility despite recent GP efforts to revoke this.

Related evidence

- Those living further away tend to be less willing to travel to the doctor.
- Availability of transport is a common reason for not attending.
- Poor organization of data has led to difficulties assessing quality of care.
- Some studies indicate the DDS visit more and give less phone advice whereas on-call GPs prescribe less and leave patients more satisfied.
- Advantages of continuing traditional on-call rotas appear small.

RELEVANT LITERATURE

Comparison of out of hours care provided by patients' own GPs and commercial deputising services: a randomised controlled trial. I) The process of care. II) The outcome of care. Cragg D and McKinley R. *BMJ* **314:** 187–189, 190–193, 1997.

Out of hours work in general practice. Iliffe S. *BMJ* **303:** 1584–1586, 1991.

57. A proprietor of a newly-built nursing home in your area asks that your practice takes over the medical care of the residents. What issues do you wish to discuss?

INTRODUCTION

The population is becoming older and the implications on workload can be considerable.

FIVE POINT PLAN

Implications for the partners

- Influencing policies on visits/prescribing/admission policy.
- Need to arrange meeting to make considered decision.
- More work in a high morbidity group.
- Any additional income for non-NHS services.
- Expanding areas of special interest or expertise amongst partners.

Practice organization

- Rearrangement of duties.
- Shared workload or nominated doctor.
- Use of over 75 years health checks and involvement of district nurse.
- Use of protocols e.g. detection of diabetics.
- Organization of prescriptions.

Issues for patients

- Need to retain personal autonomy and may prefer to keep own GP.
- Loss of continuity of care.

Decision

- Is the home sufficiently well organized and motivated to facilitate your plans to modernize care?
- Difficulty of developing clear protocols.
- May be the only practice in the area in which case there may be no choice.
- Otherwise need to avoid offending other practices in the neighbourhood.

Wider issues

- Increasingly elderly demographic.
- Evidence suggests lack of organized medical input and sub-optimal care in nursing homes.
- Underdiagnosis of problems that can be alleviated.
- Assist nursing home policies and advise on new recommendations.
- Availability of specialist community nurses to help improve care.

RELEVANT LITERATURE

Managing diabetes in residential and nursing homes. Tattersall R, *et al.*, *BMJ* **316:** 89, 1998.

58. Is an elderly health check of use? Discuss the pros and cons with respect to the relevant evidence.

INTRODUCTION

The 1990 GP contract requires GPs under their terms of services to perform an annual check in several predetermined areas of function in the over 75 age group.

It asked the GP to appraise home environment, social network (relationships, lifestyle), mobility, mental skills, senses of hearing and vision, continence, general function and use of medication. The initiative was poorly implemented and widely criticised for a lack of solid evidence.

The end points of functional improvement are soft and benefit is difficult to define and to research.

PROS

- It has been shown that 93% of patients think it a good idea.
- Admission rates fall and morale improves.
- Problems of continence, hearing, eyesight, mobility and depression are readily picked up.
- Review of medication is likely to be of value when 1 in 4 over 75 year olds have more than 30 prescriptions per year. Drug toxicity is common and underdiagnosed.
- Can be delegated to a practice nurse.
- Most elderly attend the GP each year anyway allowing opportunistic screening.
- Good evidence supports the detection and treatment of hypertension.
- Evidence suggests that the non-consulters are likely to be well.
- One study showed detection of a problem in 44% and action taken in 82% of those. (The commonest problems were physical rather than social.)

CONS

- Can cause a significant increased workload.
- The evidence for vast unmet need is not compelling.
- No clear protocol exists.
- No clear cost–benefit analysis is available.
- There is always a psychological cost to screening e.g. anxiety.
- Increases demand for services e.g. occupational therapy, chiropody, audiology.
- Only 7% of doctors find it of value.
- Can cause significantly increased referral rates with no clear reduction in morbidity.
- Uncovering new need seems to be at odds with an increasing tendency to ration.
- No benefit has been shown in screening for skin cancer, breast cancer, thyroid disease in the over 75 year olds.
- Urgent review and standardization is needed to allow audit.

RELEVANT LITERATURE

Screening elderly people: a review of the literature in the light of the new general practitioner contract. Perkins ER, *BJGP* **41:** 382–385, 1991.

Assessment of elderly people in general practice. Iliffe S, *BJGP* **41:** 9–15, 1991.

Health checks for people over 75. Harris A. *BMJ* **305:** 599–600, 1992.

59. Jane Douglas is an intelligent 14 year old patient of yours whom you have been seeing regularly for the last 6 months. The specialist you referred her to has diagnosed chronic fatigue syndrome. What are the implications of this?

INTRODUCTION

Chronic fatigue syndrome, known in the past as ME (myalgic encephalomyelitis), is a crippling but poorly understood disease. It is a complex mix of cerebral dysfunction and trigger factors and the GP may be the key professional involved in the management.

FIVE POINT PLAN

Issues for the patient

- Explore patients' attitudes/understanding of diagnosis.
- Know what advice has been given.
- Consider the psychological effects of the condition and the label.
- The patient must be involved in all plans and decisions.

Issues for the doctor

- Recognize personal knowledge and opinions on the condition.
- Encourage self-esteem in the patient.
- Non-dismissive and flexible approach.
- Show support for the family.
- Recognize feeling of impotence.
- Communication skills to allow mutual acceptance of the situation.

Educational and social issues

- Ongoing education is vital in a fourteen year old so encourage early return to school.
- Need for liaison with school/may need an educational tutor.
- Encourage development of personal education plan.
- Consider the major psychological and practical impact on the family.

Diagnosis and treatment

- Careful investigation and diagnosis to exclude underlying causes with further specialist help if necessary.
- Awareness of the three main criteria: exercise-induced fatigue, psychoneurological disturbance (e.g. loss of concentration and memory) and fluctuations in severity of symptoms for longer than 6 months.
- Advise on pattern of disease and benefits available.
- Advise on rest and light exercise, intolerance of alcohol.
- Consider medication cautiously.
- Monitor for anxiety and depression.
- Use a joint educational and medical programme.

Follow-up

- Help from other agencies but encourage support from non-sufferers too.

- Further support and information i.e. ME association, Action for ME, can fulfil a collaborative care model.
- Involvement in multidisciplinary team case review.
- Keep up-to-date with new treatments.

RELEVANT LITERATURE

The chronic fatigue syndrome: what do we know? Thomas PK, *BMJ* **306:** 1557–1558, 1993

60. A 45 year old airline pilot presents with epigastric pain waking him at night. You feel that he is probably suffering from a peptic ulcer. On what do you base your management?

INTRODUCTION

Helicobacter pylori is present in 90% of duodenal ulcers and 70% of gastric ulcers.

FIVE POINT PLAN

Issues for the patient

- Presence of other symptoms.
- Work-related stress.
- Lifestyle may further provoke symptoms.
- Impact on work.

Consultation issues

- Focused history e.g. presence of erosive drugs.
- Opportunistic health promotion e.g. smoking, alcohol.
- Appropriate examination e.g. melaena.

Diagnosis

- Consider investigation – is there need for urgent endoscopy?
- Consider need to diagnose *Helicobacter pylori* or treat it empirically.

Treatment options

- Address fears.
- Management of stress.
- Are local guidelines for management of such symptoms available?
- Symptomatic treatment or blind anti-ulcer treatment or investigate?
- Discuss your opinion on the use of medication and different pharmacological approaches before endoscopy.
- Need to document eradication of *Helicobacter*.
- Implications for long-term therapy.
- Advice on flying – refer to occupational health advisor.

Wider issues and related evidence

- Practice protocol can be developed.
- Resource implications limit availability of endoscopy and out-patient appointments.
- Cost–benefit of eradication therapy: PPI plus standard triple therapy is more effective than H2 blockers for proven DU.
- Of the testing methods (urea breath test, biopsy, and serology), serology is the easiest and least expensive. All are equivalent diagnostically.

RELEVANT LITERATURE

Challenges in managing dyspepsia in general practice. Agreus L, *et al., BMJ* **315:** 1284–1288, 1997.

Role of helicobacter in gastrointestinal disease: implications for primary care of a revolution in management of dyspepsia. Delaney B, *BJGP* **45:** 489–494, 1995.

Helicobacter and gastric cancer. *DTB* **36(8)**, 1998.

61. You are called by the parents of a 14 year old girl who have found their daughter intoxicated with alcohol and admitting to taking 35 of her contraceptive pills. She has left a note saying she wanted to die. What is your plan of management?

INTRODUCTION

Suicide attempts often present to Accident and Emergency but this case may be more sensitively handled initially by the GP.

FIVE POINT PLAN

Patient's agenda

- What effect did she feel the tablets would have?
- Explore feelings and events leading up to it.
- May have deeper issues e.g. physical or sexual abuse.

Doctor's agenda

- Sensitive consultation skills to achieve rapport.
- See patient alone and together with parents.
- Ask about any sexual history.
- Awareness that the overdose will do little harm.
- Concern over the leaving of a note.
- Is there sign of a major depression or genuine suicidal intent?
- Assess family's ability to cope.

Family issues

- Discuss recent behaviour and course of events.
- Discuss changes in family dynamics and home circumstances.
- Would she have known that the parents would be home soon?

Legal/confidentiality issues

- Ensure some degree of confidentiality.
- Illegality of underage sex (if relevant).
- Under 16 year olds with sufficient intelligence and understanding can consent to medical treatment but refusal to consent is not legally binding if the parents give consent in the child's best interests.

Management

- Discussion with Poisons Unit if unsure of toxicity.
- Is there depression sufficient to need acute admission (rare)?
- Discuss strategies with family.
- Gain agreement with all about the way forward.
- Arrange early review appointment.
- Consider later referral if needed.

RELEVANT LITERATURE

Managing self-harm: the legal issues. *DTB* **35(6)**, 1997.
Management of deliberate self-harm in general practice: a qualitative study. Prasad LR, *et al., BJGP* **49:** 721–724, 1999.

62. You review the treatment of epileptics in your practice. What models of care might be used? Give examples of how this would help your delivery of primary care?

POSSIBLE MODELS

Some form of jointly managed shared care between GPs and hospital consultants with enhanced exchange of information is widely seen as the preferred model.
Examples include the following:

- Community clinics where specialist runs a clinic in the practice.
- Basic model where letters are sent after attendances but also regular updates are sent so exceeding the normal communication.
- Liaison meetings where a case review meeting with all carers to decide on joint management plan.
- Co-operation cards carried by patient between carers and documenting an agreed data set.
- Computer assisted care where all information is inputted to hospital computer with updates and advice sent in standardized way back to GP.
- Electronic mail – common database with multiple entry and access points available to all members of the team.

EXAMPLES OF APPLYING CARE – FIVE POINT PLAN

Initiating treatment

- Local protocols can facilitate diagnosis and referral.
- Telephone/email contact with specialist can allow initiation of drug therapy.

Continuing care

- Agreed datasets for chronic management can simplify treatment.
- Updated computer records allow at-a-glance assessment of control.
- Clinical nurse specialists can help liaise with medical team.

Information and advice

- Patient-held records can be a personal source of information and encourage problem ownership.
- Patient protocols can advise on how to access help.
- Agreed written advice on driving, use of seizure diary, free prescriptions, personal drug regime, avoidance of triggers, contraception etc.

Involvement of other agencies

- District-based workers can continue education and support.
- Clinical nurse can co-ordinate with schools, employers, family and carers.
- Patient organizations (e.g. National Society for Epilepsy, British Epilepsy Association) offer information and support.

Audit

- Allows combination of practice-based, district-based and hospital-based audit.

RELEVANT LITERATURE

Diagnosis and management of epilepsy in adults – a national clinical guideline. *Scottish Intercollegiate Guidelines Network,* Nov 1997.

63. Outline the barriers to the prevention of disease that face the general practitioner.

INTRODUCTION

In all disease the goal is prevention but unfortunately despite recognizing its importance we are not very good at it. Several factors contribute to this.

FIVE POINT PLAN

Logistical barriers

- Often it is not the top priority of a consultation.
- Practice organization e.g. lack of computerized records may not allow easy identification of problems.
- Lack of personal knowledge of clear cost-efficient strategies.
- Lack of time to initiate new strategies.

Motivational barriers

- Crisis-led work patterns do not allow strategic planning.
- Feeling of overwork.
- Lack of interest in the repetitive tasks that such work requires.
- Underuse of delegation to other care team members.
- Patients' lack of knowledge of available strategies.
- Patients' fear of the doctor and reticence of men to consult.

Psychological/cognitive barriers

- As doctors we may overrate our own success on prevention.
- Risk-taking ethos may be a predominant frame of mind for stoic patients.
- Some find policies akin to badgering and incite anger and rejection.
- Lack of attachment – the "it won't happen to me" philosophy.

Financial barriers

- Practice may have to pay extra staff for additional programmes.
- Rationing in all its forms e.g. prescription charges, lack of availability of appointments.
- Lack of central funding and finance initiatives.

Political/ethical barriers

- Changing political agendas and priorities.
- Policies influenced by pressure groups.
- Ethical limitations e.g. cannot make child benefit available only to those who receive whooping cough vaccine. However, in recent years smokers have been refused major vascular operations.

64. Linda is a 22 year old patient of yours with schizophrenia who has recently been discharged from hospital. She has a history of self-harm and lives with her mother. She tells you that she and her boyfriend are intending to have a child. What areas should be considered in her ongoing care?

INTRODUCTION

Schizophrenia affects one person in 100 at some stage in life. Frequently, incomplete recovery and further relapses are the rule.

FIVE POINT PLAN

Issues related to pregnancy

- Is this a sign of impending crisis? Is she already pregnant?
- Explore reasons for wanting children.
- Need for psychosexual help.
- Ability to function as parent may be impaired.
- Issue of drug treatment during pregnancy requires specialist advice.
- Need for explanation of risks following childbirth.
- Risk of offspring suffering from schizophrenia rises to 40% if both parents are sufferers.

Family issues

- Strong support for mother is needed e.g. support groups.
- Assess needs of other family members.
- If there are high levels of criticism, hostility and over-involvement then patients are three or four times more likely to relapse.
- Prevent psychological problems in family and assess need for family intervention programme.

Treatment and maintenance

- Prepared plan for early treatment and crisis intervention.
- Opportunistically look for signs of stress, tension, relapse or depression.
- Medical monitoring of drug levels.
- Measures to minimize the risk of losing contact.
- Facilitate ongoing education and support for patients.

Multidisciplinary care

- Good relationship with key worker e.g. community psychiatric nurse.
- Need for reliable record-keeping/joint database/regular reviews.
- Assessment of required level of supervision and a clear individual care plan.
- Clear local policies on GP versus hospital responsibilities e.g. criteria for admission.
- Awareness of Mental Health Act.
- Need for additional advice/referral/specific interventions.

Social issues

- Assessment of issues of daily living and driving restrictions.
- Availability of allowances e.g. disability benefit.
- Encouragement to maintain function, work and fulfil useful role.
- Availability of voluntary agencies e.g. SANE, MIND.

65. How would you go about organizing an influenza vaccination programme in your practice?

INTRODUCTION

In the debilitated, influenza can kill. In pandemics, millions have died. Each year it is the GPs responsibility to offer immunization to a defined high-risk population.

FOUR POINT PLAN

Planning

- Look at previous year's vaccine numbers.
- Assess any differences e.g. building of a new residential home in the area.
- Identify local and national guidance on who is at risk.
- Consider the buying power of a local consortium.
- Choice of vaccine.
- Order sufficient vaccine.
- Awareness of the requirements e.g. capacity to ensure cold chain.

Targeting

- Suitable advertising e.g. poster in waiting room.
- Use of computer to identify high-risk groups e.g. chronic heart disease, diabetes, age-sex register.
- Personal invitation by letter or telephone.
- Policy for contacting defaulters.
- Liaison with residential and nursing homes.

Protocol for giving the vaccine

- Dedicated clinic or opportunistic.
- Awareness of contraindications e.g. egg allergy.
- Information for patients e.g. what to do in the event of a reaction.
- Delegation to suitably trained nurse.
- Opportunistically offer further health assessment.
- Clear record keeping – date, batch etc.

Follow-up

- Medical review for any adverse reaction. (Consider yellow card reporting.)
- Claims for dispensing fees.

66. David Campbell is a 45 year old businessman with a recent conviction for drink-driving. He has been sent by his wife for treatment for his alcohol problem. He admits to drinking a bottle of whisky a day. How would you progress?

INTRODUCTION

One in 20 people are dependent on alcohol. An empathetic and holistic approach considering psychological, physical, work and social issues is necessary to initiate successful treatment.

FIVE POINT PLAN

Issues for the patient

- Assess personal motivation for change.
- Explore own agenda and concerns.
- Consequences for work and home life.
- Explore understanding of the effects of alcohol and expectation of GP.

Issues for the doctor

- Empathetic consultation skills and non-judgmental approach.
- Assess mental status and risk of self-harm.
- Full drinking history.
- Examination targeted to alcohol-related illness.
- Appropriate investigations e.g. FBC, LFT etc.
- Knowledge of methods of safe withdrawal in community.

Relationship issues

- Effects on relationships and family life including sexual function.
- Underlying financial problems.
- Support will be essential for successful detoxification.

Treatment options

- Decide on goals e.g. decision to advise reduction or abstention.
- AUDIT (Alcohol Use Disorders Identification Test – W.H.O.) helps identify dependence.
- Five to 10 minutes of brief advice on the risks the patient is taking is an effective medication.
- Lifestyle advice to redirect social patterns.
- Agree plan with patient.
- Community treatment or referral to specialist centre.
- If referred, share monitoring of progress.

Further help

- Ongoing primary care support and follow-up as motivation allows.
- Vigilance for anxiety and depression.
- Advice what to do in likely event of relapse.
- Advise attending support groups such as Alcoholic's Anonymous.

- Additional specialist help e.g. psychiatrist or counsellor.
- Medications such as acamprosate have been shown to reduce desire to drink.

RELEVANT LITERATURE

Tackling alcohol abuse. Deehan A, *et al., BJGP* **48:** 1779–1782, 1998.

67. Julie is a 20 year old with a pregnancy of 9 weeks. She comes to you and says that she thinks she wants an abortion. How do you proceed?

INTRODUCTION

One in three pregnancies worldwide ends in termination with 150 000 in Britain annually. A sensitive approach is required to minimize psychological trauma.

FIVE POINT PLAN

Patient's agenda

- Explore concerns and personal circumstances.
- Understanding of why pregnancy occurred.
- Has there been discussion with the father?

Doctor's agenda

- May be happier seen with friend, partner or relative.
- Empathetic communication skills and written information.
- Need to confirm pregnancy.
- Arrange review to rediscuss or fast referral to specialist.
- If own moral code does not allow agreement, refer to colleague.

Information and counselling

- Guarantee confidentiality.
- Enable patient's decision by sharing options and alternatives e.g. adoption, continuing with pregnancy.
- Allow time to come to decision that she will least regret.
- Counselling may be carried out by GP or other advisory service e.g. British Pregnancy Advisory Service.
- Discuss financial help for continuing pregnancy.
- Details of termination methods and complications.
- Opportunistic care e.g. smoking, chlamydia screening.

Follow-up

- Decision on future contraception.
- Exclude complications.
- Encourage ventilation of feelings and assess for maladaption.

Wider issues

- Legally, terminations are carried out after two doctors agree in accordance with the 1967 Abortion Act amended by the Human Fertilisation & Embryology Act for one of five possible reasons.
- Local policy varies on the availability of termination on the NHS. Countrywide about half are done privately but there is massive regional variability.

68. Discuss the GPs role in the management of domestic violence.

INTRODUCTION

Domestic violence is the term used to describe a type of physical, emotional and mental abuse. As many as one in four women suffer violence at the hands of men with whom they share an intimate relationship. It crosses all social and ethnic boundaries.

FIVE POINT PLAN

Issues for the patient

- Use of a "calling card" hiding the real agenda.
- Sense of helplessness and low self-esteem leading to psychiatric morbidity.
- Fears of social and economic isolation.
- Without intervention problem usually escalates.

Index of suspicion

- Consider diagnosis based on clues in the consultation.
- Clues include unexplained injuries, inconsistencies, evasive manner, clues in the notes, persistent presentation with minor illness, history of attempted suicide, depression.

Issues for the doctor

- Doctor's lack of confidence in ability to help.
- Awareness of local services.
- Emphasize confidentiality.
- Ask! Be sensitive and non-threatening. Evidence shows that those affected want to be asked directly and those not affected do not mind being asked.
- Assess present situation gathering information about sources of support, living circumstances etc.
- Clear documentation including verbatim history, times and dates, description of injuries (consider photographs).

Care of children

- Mother may fear disturbance of relationship with father.
- Are children in danger?
- Consider referral to social services if at risk. Try to get patient's consent but you must act in their best interests. Further advice may be useful.

Advise and plan

- Explanation that domestic violence is illegal and that the patient is a victim.
- Explain physical and emotional consequences of ongoing abuse.

- Provide written information on legal options and help from police domestic violence units, Womens' Aid National Helpline, local authority social service and housing departments. Offer help in contacting them.
- Empower patient's decision by encouraging autonomy and self determination.
- Be available to discuss again.

RELEVANT LITERATURE

Domestic violence: the general practitioner's role. Heath I. *RCGP* 1997.

69. What are the pros and cons of GPs performing intrapartum care?

INTRODUCTION

A clearly researched comparison of home versus hospital care would be useful. Unfortunately, an ethical randomized controlled trial would be a contradiction in terms.

ANSWER PLAN

Pros

- Expression of maternal preference is part of a healthy birth process.
- About 8% of women would prefer a home birth.
- Maternity services should be sensitive to the requirements of the local population.
- New models of care may allow GP input to break through practice boundaries.
- Care can be organized in GP obstetric units to best use local skills.
- Community midwives are increasingly keen on home delivery.
- Home deliveries are cheaper than in hospital.
- More continuity of care may produce a better birth experience.
- GPs retain a duty of care in an emergency and may have to attend unplanned delivery.

Evidence has shown

- Routine electronic monitoring during labour does not reduce perinatal mortality but does increase the chance of caesarean section.
- Studies show that planned home births have a very low mortality rate (but data excludes higher risk births by definition).
- One study showed 65% of consultant obstetricians feel it is safe for low-risk pregnancies (but 35% felt it was less safe than hospital care).
- Low-risk pregnancies may have little or no benefit from obstetrician-led shared care compared to routine GP/midwife care.

Cons

- Doctors have become deskilled in this area and the vast majority do not wish to be involved in deliveries.
- GP registrars have little experience of normal delivery.
- Delivery rates are insufficient to maintain skills even if interested. This presents problems of ongoing training.
- Lack of personal round-the-clock availability of the modern GP.
- Difficulty of incorporating disruption to work responsibilities.
- The GP taking responsibility for delivery exempts himself from the Bolam test as it usually applies to GPs. He legally undertakes to provide a standard of care equivalent not to other GPs but to an obstetrician.
- Most practices are not equipped for home deliveries.
- Things go wrong very quickly.

Evidence has shown

- Despite a very low perinatal mortality rate, estimations suggest that 42% of deaths in healthy babies could be prevented by better care.
- Babies at high risk are safer under specialist hospital obstetric care (and a normal delivery is a retrospective diagnosis!).
- Only 75% of GPs currently providing intrapartum care want to continue.

RELEVANT LITERATURE

Changing Childbirth. *Dept of Health* 1993.

Should obstetricians see women with normal pregnancies? A multicentre RCT of routine antenatal care by general practitioners and midwives compared with shared care led by obstetricians. Tucker JS, *et al., BMJ* **312:** 554–559, 1996.

Opinions of consultant obstetricians in the Northern region regarding the provision of intrapartum care by GPs. Frain JPJ, *et al., BJGP* **46:** 611–612, 1996.

Babies deaths linked to sub-optimal care. Dillner L, *BMJ* **310:** 757, 1995.

70. What are the advantages and disadvantages of walk-in centres?

INTRODUCTION

Walk-in centres are a new and separately funded government-backed initiative.

ADVANTAGES

- Convenient access for non-urgent appointments.
- A response to changing needs.
- May be popular with patients.
- Designed to work as an additional service rather than replace family doctors.
- Another channel to help manage minor illness.
- May identify unmet need and help to redress inequalities in healthcare delivery.
- Separate funding avoids competition.
- Opportunity to experiment with new health care delivery models.
- Possibility of wider range of treatments available.
- May release some of the pressure on the primary care team.

DISADVANTAGES

- A worthwhile development or a populist whim?
- Has there been a widely considered discussion of alternatives and an agreed plan sharply focuses to meet needs?
- Lack of good communication network between walk-in centres and practices.
- Fragmentation: lack of continuity of care.
- Duplication: pharmacists already provide a limited drop-in service.
- Could undermine the traditional values of primary care and be the first step towards the destruction of a unique service.
- Failure to provide adequate follow-up.
- Records can no longer be comprehensive.
- Services may well compete.
- Registrar training focuses on the essential need to understand underlying beliefs and social aspects. Holistic care is largely denied in the walk-in centre.
- Funding may be better spent improving GP care e.g. patient education, smaller lists, more effective use of skill mix.
- Who will man the centres? – a practice nurse without a medical qualification or a doctor who may be much needed to improve care in a busy practice?
- Patients need to be educated to select the right service.
- Need to audit and show benefit.
- Reduced ability to tackle repeat prescribing, referrals.
- Reduced awareness of allergies, knowledge of interactions.

- Putting a "patch" on primary care with an add-on service is not the way to improve it when major structural changes are needed and on the way.
- May increase the workload of the GP to the relief of A&E departments.
- Twenty years of experience from Canada still gives widespread concerns. The patients are seen superficially and go on to see their own doctor in addition.

RELEVANT LITERATURE

Should PCGs back walk-in centres? *Medeconomics* July 1999.

THE 4 MINUTE VIVAS

The emphasis of the oral part of the MRCGP is aimed at these three broad areas:

- Communication
- Professional values
- Personal growth

The communication area encompasses verbal and non-verbal communication techniques, skills for effective information transfer, principles of communication and consultation models.
Professional values cover moral and ethical issues, patient autonomy, medico-legal issues, flexibility and tolerance, implications of styles of practice, roles of health professionals and cultural and social factors.
Personal and professional growth focuses on continuing professional development, self-appraisal and evaluation, stress awareness and management, burnout and change management.

These are tested in four contexts:

- Care of the patient
- Working with colleagues
- The social role of general practice
- Personal responsibilities for care, decisions and outcomes

The oral comprises two similarly-structured 20 minute sessions each with a pair of examiners. An observer may also be present. The sessions are separated by a 5 minute break for changing tables. The examiners will have no knowledge of your performance in the other modules of the exam or the other viva. They will have planned your orals in advance so as to ensure examination of different topics and they will take it in turns to question you. They will typically cover five questions per table. The watch on the desk counts 4 minutes per question. One speaks while the other marks.

The orals exist to test areas not tested in the other modules.
They will assess your decision-making processes in day-to-day practice and your ability to justify your conclusions in the face of critical challenge from the examiners. It is important to demonstrate this process in your dialogue. The journey is often more highly weighted in the marking scheme than the destination.

The questions can cover any area of practice and clinical scenarios are often used to illustrate not particular clinical solutions but your approach to them e.g. use of consultation skills. It is sometimes useful to remember the same approach you have used in the written paper. This adds value to the idea of doing the viva module in the same sitting of the exam as the written paper. You must know what is in current journals in particular the *BMJ* and *BJGP* and in the media generally. A broad preparation for the written paper will ensure you stumble across most of the topics and issues that are current. Consider your position in advance on these issues and practice presenting them in a concise, well structured and confident manner.

Listening carefully to the question will help you realize what areas and contexts are being tested and allow you to target the structure of your answer.
Practice of the technique is essential because it is surprisingly difficult to give good viva answers even on subjects with which you are familiar. Because it can be difficult for many to get good practice, I have included a series of sample questions to try. The first group of questions are supplemented by notes which highlight important points. Additional questions are included to allow further practice.

As classical dilemmas frequently appear in both the written and oral parts of the MRCGP, a further list of these is included for you to consider.

The viva questions are best practised in a semi-formal environment with a friend or colleague asking the questions. This enables interactivity and is more realistic, allowing the questioner to go off at a tangent. It also permits the process of challenging your opinions. If you have a strong opinion do not be afraid to express it but it will need balance to show you have considered the other options. As long as you can justify your thoughts the examiners will appreciate a bit of enthusiasm and it may well benefit you. Remember to refer to evidence in your answers if it seems relevant.

Difficulty will increase as you are challenged and you may reach a point beyond which you cannot go. It will feel uncomfortable but do not be put off. It is all part of assessing your limits.

Typically, the initial question may set the scene before the follow up questions focus on the area the examiners are testing. I have therefore presented the sample questions in the following format:

STEM QUESTION
- FOLLOW-UP QUESTION
- FOLLOW-UP QUESTION

The pace is rapid. Say your piece and then stop. The examiners will be ready with their follow-up.

PRACTICE VIVAS WITH NOTES

VIVA TABLE 1

Question 1

- What might be the value of patient-held records?
- What would be the advantages for patients?
- How might it work in practice?

Notes

Patient-held records (PHRs) may become a reality in the NHS. Early research has identified practical models which patients find acceptable and desirable. They feel more involved, better informed and it can improve co-ordination of care. They can correct errors in the record. They require the co-operation of all the medical practitioners involved in a patient's care and careful practice organization to ensure PHRs are updated.

One model would involve the replacement of all practice written records with a practice computer record and a patient-held written or computer printed summary. More research is needed.

Question 2

- What do you find stressful about general practice?
- What is burnout?
- How would you avoid it in your professional life?

Notes

Burnout is a syndrome consisting of emotional exhaustion, depersonalization of others and a lack of personal accomplishment. It leads to diminished performance, fatigue and inefficiency. Stresses such as living with uncertainty, unforeseen errors, inappropriate demands, heartsink patients and unsociable hours require recognition. There is no association between age and burnout. Survival skills include good time management, development of other interests, planning breaks, forming realistic goals and keeping your sense of humour.

Question 3

- You are asked to supervise a boxing match by the local boy's club. It will be cancelled unless you do it. Will you?
- Does the BMA have any opinions about this?
- How would you feel about going against the wishes of your union?

Notes

Justify your choice of answer. Will you let your own opinions on contact sport predominate? Be aware that the BMA has campaigned for some time for stricter legal regulation of boxing. It is the only sport where injury is a result of correct performance. Assault can be legal because of consent e.g. in the case of surgery, but some feel that standards are inadequate and that could leave a participating doctor open to be sued for negligence.

Question 4

- What techniques do you know for breaking upsetting news?
- What else can you do to make it easier?
- How does it affect you?

Notes

This question may be dressed up in a scenario making it easier to overlook the principles. Not all bad news is gravely bad but you must be sensitive to the impact on the patient. Be prepared, check what is known, fire a warning shot, and explain slowly and clearly. Allow the patient to respond and empathize. The use of contact, diagrams and tape-recording the session to play back later can make a world of difference.

Question 5

- What do you understand by "morbidity registers"?
- What are the implications for running your practice?
- Will patient care improve as a result?

Notes

The change from the treatment of individuals to the management of populations requires increasingly well-kept registers of patients with particular medical conditions. All practice staff need to understand the purpose and importance of the registers and their commitment ensured to keep them up-to-date. A plan identifying sources of information and a programme for updating the register needs to be agreed. It may be that remuneration will extend more widely than the current areas of asthma and diabetes.

Population strategies and implementation of guidelines is difficult unless the relevant patients are found and given the opportunity of the latest recognized treatment. Audit will exact increasing standards and practices will be required to illustrate performance.

VIVA TABLE 2

Question 1

- What makes consultations go wrong?
- How can you make sure patients understand what you tell them?
- What are the consequences of getting it wrong?

Notes

Poor consultation technique results in a poor doctor–patient relationship. The fault may lie in simple matters such as the use of language on both sides of the consultation or more complex issues relating to underlying health beliefs which require time to explore but are crucial to a good relationship that ensures trust and compliance. Daily pressures and prejudices affect how we operate. Use of open posture, eye contact, an attentive pose and other techniques to feign interest help. Checking understanding and backing it up with written information helps. The effective GP must have a career-long dedication to minimize the many negative influences upon clear communication. Complaints and even more stress are the rewards for failure.

Question 2

- Why do you want to be a GP?
- What will you do if you do not feel you know enough?
- Why should you bother keeping up to date?

Notes

Rewarding aspects of practice include independence, holistic nature of care, the challenge of problem-solving, opportunity to work flexibly, continuity of care, status in society and financial rewards. Lifelong self-directed learning requires targeting of weak areas. These may be identified during everyday work or discussion groups to build up a personal learning plan. Reading, listening to patients and teaching are all ways of identifying the plan. The GMC's Good Medical Practice requires you to keep your knowledge and skills up-to-date. The revalidation of GP principles will incorporate aspects of an educational portfolio and the process should result in increased professional satisfaction.

Question 3

- A patient comes to see you asking you to recommend a cosmetic surgeon for breast enhancement. What are your feelings about this?
- What is the GPs role in this area?
- Does cosmetic surgery do any good psychologically?

Notes

In most areas, cosmetic surgery is not available on the NHS in the absence of significant psychological morbidity. Many now self-refer to clinics but some will approach their GP for advice on a safe approach. This gives an opportunity to explore underlying personal/relationship issues and discuss

any risks and complications and encourage realistic expectations. If the decision to continue is made, a referral letter summarizing the patient's background will be of use.
The evidence shows a psychological benefit to cosmetic surgery, particularly in the short-term.
To help the patient choose a plastic surgeon, contact the British Association of Aesthetic Plastic Surgeons to identify well qualified and experienced surgeons. The GMC also holds a record of qualified specialists who have undergone specialist training. A phone check can be done on a named surgeon. Some surgeons in cosmetic clinics may lack expertise.

Question 4

- Whose model of the consultation do you follow?
- How do you find someone's hidden agenda?
- What is meant by the "infinite potential" of the consultation?

Notes

Discuss your favourite model. For example, Pendleton's seven point plan aims to define reasons for attendance, consider other problems, choose appropriate action with the patient, achieve shared understanding, involve patient in management and encourage to accept responsibility, use time and resources appropriately and establish or maintain a relationship to help achieve other tasks.

Question 5

- Mrs. Carr is an anxious patient of yours. She returns to see you a week after you visited her at home and prescribed penicillin for her tonsillitis. She says she is no better despite finishing the course and presents the empty bottle of tablets. The label says penicillamine instead of penicillin. She is otherwise well. What is your reaction?
- Do you tell her the truth?
- What is significant event analysis?

Notes

Horror! Informing her will be distressing for both parties although you have not yet confirmed whether the error is yours or the pharmacist's. A visit to see him is necessary. If the mistake was entirely trivial with no potential consequences, you may choose not to inform her but you have a duty as a doctor to tell the truth especially as the drug is potentially toxic. You could only justify non-disclosure if you felt giving the information would be likely to cause serious harm to her mental health. A little anxiety does not qualify. You may face anger and distress but hopefully honesty will strengthen the doctor–patient relationship. A critical incident review should ensure that it does not happen again.

VIVA TABLE 3

Question 1

- How would you react to a request by a patient to see her notes?
- Are there any circumstances in which you would refuse?
- What if she disagrees with some of the non-medical facts and requests you change them?

Notes

It is useful to understand the reasons. A review of the notes may allow this to be ascertained. A written application by the patient is necessary and access given within 40 days. Third party information should be removed unless the third party gives consent. The Access to Health Records Act 1990 allows access to notes written after 1 November 1991 and requires a health professional explain anything not understood. No legal right exists to notes written before that time but as a general principle they should be released. Outright refusal can only take place if the release is believed to lead to the possibility of serious harm to the physical or mental health of any party. This decision can be challenged in the courts. Non-medical facts can be corrected but if they are believed accurate, the best approach is to add a comment about the patient's objection.

Question 2

- What are the problems with one patient consulting on behalf of another?
- How can you maintain confidentiality?
- What will be your approach to the consultation?

Notes

Third party consultations present difficult issues of ethics and confidentiality. The scenario (e.g. mother about daughter or wife about husband) will influence the specifics of your answer. Focus on the underlying reasons and try to identify which party needs most help. Is this a sign of disharmony in the relationship, a personal cry for help or genuine concern for the third party? Avoid a breach of confidentiality and act in the best interests of your patient. You may need to take steps to get permission to divulge information. Your main aim is to promote communication between the two parties.

Question 3

- How do patients put doctors at risk?
- How might doctors put their patients at risk?
- What steps will you take to avoid this throughout your career?

Notes

Doctors may inadvertently put patients at risk due to physical or mental illness, poor decision-making and lack of knowledge or competence. Efforts to optimize skills such as listening, identifying goals and giving information in a structured manner along with principles of lifelong learning, and learning from mistakes help to minimize this.

Question 4

- An 18 year old man presents a health report for you to fill out for the army. He asks you to omit the diagnosis of asthma that he was given at the age of 5 as he has not used inhalers for the last 7 years. What is your response?
- If you do not do as he asks he will be refused a chance in the army. You know his educational background is poor and unemployment in his peers is high. How does this affect your opinion?
- Would you ever lie for a patient?

Notes

You have an obligation to tell the truth or be open to litigation. The stress of a new physical challenge could provoke some of his previous symptoms. A doctor's signature must be a guarantee of probity.

Your patient is putting you in an unreasonable position and perhaps leaving you open to litigation and a claim of conspiracy to defraud as well as serious professional misconduct. The information is relevant and there is a moral right to reveal it. Full disclosure regarding the history of the illness and lack of symptoms is a safe approach and can allow the army's medical adviser to make an appropriate decision. Lack of honesty at this stage will not find favour with an employer.

The patient has a right to read the report before it is sent under the Access to Medical Records Act 1988. If the patient delivers the report there is even an opportunity for him to change it without your knowledge!

Question 5

- Do you agree to see pharmaceutical representatives?
- How do you view their promotional material?
- What would influence you to prescribe a new drug?

Notes

Drug reps can be seen as the bearer of free pens (and a pleasant source of relief for a GP) or a pest. Research suggests they are still the source of much information on drugs for GPs but this must be looked at carefully using the principles of critical appraisal. The developments that the companies finance bring an array of new advances to improve patient care. Doctors can be divided into innovators, moderators and procrastinators depending on their keenness to try new drugs. Factors which influence the confidence to use new drugs include cost, perceived benefit, presence of guidelines and use of the drugs by specialists.

VIVA TABLE 4

Question 1

- You receive a request for a statin from an insistent female patient who does not fit the guidelines for treatment. How would this make you feel?
- Outline your approach.
- What if she perseveres?

Notes

Patients have access to a lot of information but can fail to apply it appropriately. She requires an understanding approach and reassurance that the benefit of the medication is not considered sufficient in her circumstances. Explanation of the underlying evidence for your decision and the lifelong commitment of treatment and side effects may help. There is no obligation to prescribe on demand and you may face criticism in the face of an adverse event. You are a gatekeeper for resources and it is not possible or desirable to treat everybody. It may initiate similar requests.
If you fail to reassure her, you would have to justify a decision to prescribe even if it was done as a private prescription which she would find expensive.

Question 2

- What do you feel when a patient brings a list of problems to the consultation?
- Are there any ways to turn this to your advantage?
- What communication skills will help you here?

Notes

Admit your genuine feelings and recognize your frustrations. It is known that the majority of patients have more than one problem for each appointment. Learn to accept this. It can help to organize workload. Assume a list is present and listen carefully without interrupting initially. It is an important task to discover this list early in the consultation and prioritize it. Evidence shows that doctors interrupt their patients' opening statements after an average of 18 seconds. And that allowing continuation elicits more concerns. The first problem may not be the most important. Be alert for non-verbal clues and reflect these back. Ask early on "is there anything else?" and negotiate an agenda for the consultation. Seeking out problems will save time in the long run.

Question 3

- What are the strengths and weaknesses of practice-based learning?
- How does so-called evidence-based research help you learn?
- How will you avoid getting stale as a GP?

Notes

Practice-based learning can share knowledge and insights, allow effective use of resources and strengthen the team but other members of the primary

health care team may not have the same educational needs. Plans must always be tailored. Methods of broadening your outlook include teaching, extra courses, clinical assistantships and A&E sessions.
Evidence provides a good framework for learning and better communication with patients but it cannot be applied crudely. Results cannot be extrapolated beyond the choice of research group and many outcomes cannot be measured.

Question 4

- What is gained by removing a patient from your list?
- What would make you remove a patient from your list?
- How would you try to prevent this arising?

Notes

The RCGP's guidance on removal of patients from the practice list justifies removal in three circumstances – crime and deception e.g. deliberate lying to obtain benefit, violence including verbal abuse and distance e.g. moving out of the designated practice area. In the event of irretrievable breakdown of the doctor–patient relationship, it also recommends a careful plan of communication with colleagues and patient, a review of contributory factors, and a succession of steps to be taken to remove the patient from the list.

Question 5

- A 15 year old girl comes to you and admits she is being sexually abused by her father. She is happy to talk to you about this but requests you tell no-one as she will be leaving for college soon. What is your reaction?
- Will you break her confidence?
- Where does your duty of care lie?

Notes

Believe her and listen sympathetically. She will need advice on pregnancy and sexually transmitted disease. Do you consider her to be Gillick competent? You owe her a duty of confidentiality but have a duty of care to protect her from further abuse. You must decide if it is her best interests to disclose to a child protection team. This is best done with her consent, so make every effort. Failure to act on the information could be seen as condoning incest and child abuse. However, action may provoke a serious reaction such as running away. Other help such as counselling or sexual abuse helplines could be offered. As her GP you may well feel you need to discuss the situation with a specialist.

VIVA TABLE 5

Question 1

- When you are an employer (a GP partner), how will you get the most out of your staff?
- What is meant by staff appraisal?
- How might it affect the development of a group practice?

Notes

Staff appraisal is a widespread technique of continual review that allows review of employees' performance, goals and development. It allows a more informed relationship with the employer and clarification of duties as well as identifying problem areas and training needs. It is done in protected time and forms are available to give structure to the session. Most of all it is an opportunity to give praise. A smile and a thank you are invaluable. Use them liberally.

Question 2

- Do you think GPs should be involved with research?
- Where might you get help for an interesting research project in general practice?
- What are the implications of doing research on patients in practice?

Notes

Developing an interest in research is a healthy part of the learning process for the interested GP. Practical help and funding can be accessed via your clinical tutor, post-graduate professor, from local research groups and centrally from the RCGP. The National Group on Research and Development in Primary Care has been criticised as not being sufficiently patient-orientated. Only 22% of its research is clinical. A greater proportion in the past has been related to organization and GP attitudes.

Question 3

- Why do patients come to see a doctor?
- Why do some patients consult frequently for trivial complaints?
- What strategies do you think can tackle this?

Notes

Patients can approach a GP for many reasons such as pressure from relatives, reduced personal responsibility and to legitimize sickness as well as the need for treatment/reassurance.

Frequent attendance can represent inability to cope, poor home support, an unresolved agenda, an underlying undiagnosed problem, inappropriate health beliefs and unrealistic expectations. Doctors require good communication skills to find out which. Practices can arrange a meeting to broach the problem with strategies such as firm (rather than opportunistic) arrangements for review, agreeing to avoid unnecessary prescribing or giving too much time which may confirm the patient's expectations. A sympathetic approach incorporating the patient's own health beliefs can give confidence for self-management. Resources such as self-help groups, day centres or anxiety management courses may have a role.

Question 4

- What do you understand by clinical governance?
- What is the Commission for Health Improvement?
- What is NICE and how will it affect your practice?

Notes

Clinical governance is the initiative of the White Paper, The New NHS: Modern and Dependable. Its agenda is quality. It involves action to avoid risks, investigate adverse situations and disseminate the lessons learned. It advocates systems that ensure continuous improvement in clinical care at local level and throughout the NHS. It is a framework to improve patient care through commitment to high standards and personal and team development. It covers a wide range of areas from identifying poor performance and unacceptable variations in care to promoting lifelong learning and personal accreditation. It advocates the development of cost-effective, evidence-based clinical practice in an accountable and transparent system. The Commission for Health Improvement will oversee the quality of clinical governance and of services.
The National Institute of Clinical Excellence is to produce and disseminate clinical guidelines to promote cost-effective therapies and uniform clinical standards.

Question 5

- How can you assess quality in general practice?
- What is the role of the college in this?
- How can PACT data help?

Notes

Many variables have been used. The RCGP has taken a central role in this and its document "Recognising Quality of Care in General Practice" describes these. Single performance indicators are a crude and inaccurate assessment of quality. They include ability to reach screening targets as well as measurement of referral rates. (Does a good doctor refer more or fewer patients?).
Prescribing patterns (as shown in the Prescribing Analysis and Cost data) such as the ratio of inhaled steroids to bronchodilators have not in themselves been shown to be a good index of the overall quality of GPs. Indeed achieving a performance indicator can replace the pursuit of high quality care. Quality is indicated by the like of vocational training, possession of the MRCGP, Fellowship by Assessment and the Quality Practice Awards. Practice Accreditation and Accreditation of Personal Development continues this trend. Measuring quality requires a country-wide, multi-dimensional framework covering all areas of general practice including clinical audit. It may be that at least initially, quality markers need to be set at different levels to stimulate and encourage GPs who are on different points on the quality continuum.

VIVA TABLE 6

Question 1

- A mother comes to the surgery and says she is certain that her husband is going to take her daughter out of the country to have her circumcised. She is desperate for help; what is your advice?
- Whom might you approach for advice?
- What might a social services referral achieve?

Notes

The Prohibition of Circumcision Act 1989 makes this procedure an offence except on specific physical or mental grounds. It is child abuse and your duty is to safeguard the child's welfare. A Child Protection Investigation involving social services should take place. The aim is to clarify the facts, identify areas of concern, and grounds for concern, looking for evidence to suspect the child is at risk of significant harm and decide on where the child will be safe. It will involve a multidisciplinary team with sensitive communication skills in an open and honest discussion with all the relevant parties.

Question 2

- How can you avoid complaints?
- What action should be taken to deal with a complaint to the practice?
- What are the possible outcomes of a complaint?

Notes

A story always has two sides and frequently a breakdown of communication is to blame for complaints. The GMC requires you to give a prompt, cordial and full explanation of what went wrong and an apology if appropriate. The vast majority of complaints can be dealt with by an in-house complaints procedure. This delegates the initial action including contact with the complainant within 2 working days and facilitates appropriate investigation and action. Outcomes include referral to the health authority which may result in independent review, referral for conciliation, referral back to the practice or advice on approaching the Health Service Commissioner (or ombudsman) who can recommend a discipline committee investigation. Ultimately cases can be referred to the GMC or to an NHS tribunal.

Question 3

- Flexible ways of working such as part-time, locum or job-shares are on the increase. What impact do you think that growing trend will have?
- Would you prefer to work as a locum or a principal?
- How would you prevent isolation as a locum?

Notes

There will be implications for the practice who will wish their many duties to be shared by the others working in the practice. There may be fewer to

share the out of hours service or attend practice meetings. There is more to being a GP than coming in for two or three surgeries a week. Communication and continuity of care is further eroded. Voting rights may need discussing in the case of job-sharing principles. On the other hand, it may give an opportunity to increase the range of services a practice offers. Implications for the doctor are positive. Many feel the loss of earnings is more than outweighed by the improvements in lifestyle. Other interests, medical and non-medical can be taken up. Stress is less and part-timers show a lower rate of burnout.
The National Association of Non-principals and regional locum groups such as the North East Locum Group give a voice to the growing number of flexible workers and help outline the requirements of the locum doctor.

Question 4

- What makes you bring a patient back for review?
- Why might one partner consistently run his surgeries later than another?
- How might this be tackled in practice?

Notes

Review can be initiated by doctor or patient. It can represent lack of confidence or a need for emotional involvement or to impress a patient. The passage of time is needed to make many diagnoses and review of progress is part of good care. Stabilizing a chronic condition and offering psychological support all take time. Regular appointments may also reduce the need for emergency appointments in some patients but care should be given to avoid over-prescribing for conditions which may well not have presented to an emergency appointment. Appointments can get blocked and one doctor consistently being booked well in advance can cause practice tensions. This may be due to different expertise and case-mix. A practice meeting can air the issue. Solutions include a change in the doctor's habits, adoption of personal lists or introduction of a catch-up break to the surgery.

Question 5

- How can Health Improvement Programmes affect primary care teams?
- Are you aware of any such developments in your region?
- How might such changes be organized?

Notes

Health Improvement Programmes are the lynchpin of the government vision for the NHS. New working relationships will cross health and social divides to implement national guidelines at a local level led by the Health Authority. Other organizations such as patients, schools and employers should be involved in the process. They will address needs-assessment, resource-mapping, priorities for action, strategies for change and a framework for funding. The management cycle for all processes of change parallels that for audit. Define the problems, select the goals and outcome measures, clarify resources, implement, monitor and feedback.

VIVA TABLE 7

Question 1

- What is the LMC?
- What is the GMSC?
- Would you be interested in becoming a member later on in your career?

Notes

The local medical committee is a group of peer-elected GPs and the only true representative body of GPs working in a particular area. Members are appointed to various committees of the health authority e.g. new health promotion activities. There is an elected chairman.
The General Medical Services Committee consists of GP members elected by several LMCs and is the sole representative of all GPs in the country. It is the executive at the annual conference of LMCs. Although technically a standing committee of the BMA, membership is not a requirement. It connects the voice of local representatives to the negotiating sub-committees of the Department of Health.

Question 2

- What does the term heartsink mean to you?
- Can you think of any in your practice?
- Can you identify what sort of groups they fall into?

Notes

Heartsinks are demanding patients who consult frequently and invoke feelings of despair, anger and frustration in the primary health care team. They are typically female, over 40, demanding, frequent attenders who lack insight and suffer with a high degree of anxiety/depression. Their locus of control empowers their GP as "the powerful other", believing that the doctor is in charge of their health. Models have identified four categories – the dependent clinger, the entitled demander, the manipulative help-rejecter and the self-destructive denier (Groves, 1951).

Question 3

- What does the Royal College of General Practitioners do?
- Why do you want to be a member?
- What can you offer the college?

Notes

The work of the college includes development of policy and clinical guidelines, raising standards and quality in primary health care, providing education and training, facilitation of research, producing publications, encouraging recruitment and influencing government policy.
It produces a dedicated journal, has a comprehensive information service and advances GP-led training. It enables access to becoming a trainer or a fellow and connects like-minded people in encouraging positive aspects of practice. On the other hand, the journal has been widely criticised as dull

and lacking usefulness and there is a lack of devolved regional power. After achieving the standard, there are probably few immediate advantages for the new principal. (This would be a brave approach to take in the exam!)

Question 4

- What does the practice nurse do in your practice?
- Would you rather employ a nurse practitioner or a practice nurse?
- How do you see their roles developing in the future?

Notes

Practice nurses are a valuable member of the primary care health team. Their interests and abilities enable extension of their nursing role to other duties such as pill checks, chronic disease management, HRT clinics, assisting minor surgery and advice on health education. Nurse practitioners are a much more expensive resource with major implications for training and protocol development for the practice. They may be a triage resource or run their own lists. They see about half as many patients per hour as a GP and their competence must be guaranteed by the supervising GP. Their lack of knowledge of physiology and pharmacology and lack of experience in taking responsibility for therapy as well as lack of prescribing powers limits their role.

Question 5

- How do patients get most of their information?
- What are patient participation groups?
- How can they help the practice?

Notes

Many sources inform patients nowadays. These include doctors, other health professionals, pharmacists, family, friends, self-help groups, product inserts within medication, information leaflets, radio, television, newspaper magazines and the Internet. Much of it varies in quality and relevance to the individual.

Patient participation groups are patient representatives who can liase with a nominated GP and cover various roles. They may help deal with complaints, administer donations, aid awareness of services for patients, aid planning of developments and identify unmet needs. Their liaison with the lay member of the primary care group better represents views.

VIVA TABLE 8

Question 1

- Do you think cannabis should be prescribable?
- Many would disagree with you – what do you think concerns them?
- Do you think the BMA would support legalisation?

Notes

Cannabinoids appear to have use as a treatment for symptoms such as pain, insomnia, helping opiate withdrawal and AIDS-related problems. No deaths have been attributed to it alone but rigorous pharmaceutical research with randomized controlled trials are needed. Further work will help identify the useful cannabinoids. The BMA produced a book on the Therapeutic Uses of Cannabis in 1997. Options of limited prescribability through to the consequences of legalization may be discussed.

Question 2

- How do you filter through the sources of information that a GP receives each week?
- What article has changed your practice?
- How would you organize your practice to make sure nothing important is missed?

Notes

Identify the factors of a publication which make it interesting and relevant to you. You will need to develop strategies to keep on top of developments in a time-efficient way. You may choose to concentrate on a few sources, maintain a questioning approach, read to remember e.g. by highlighting key facts and reference the articles you keep. The value of articles written by GPs could be discussed.
The importance of practice failsafes cannot be overstated.

Question 3

- You spend 15 minutes performing an elderly driving medical on a 77 year old man. The findings are satisfactory. You present him with a bill for the BMA recommended fee. He says it was done the previous time by your partner for free. What are your feelings about this?
- What implications are there for the practice?
- How will you prevent this happening again?

Notes

After spending your time on this you are bound to feel irritation. Recognize it and step back. Clearly the patient should have been warned to expect the bill. Your partner's activities should be discussed as it has impact on the rest of the practice and he is partly responsible for jeopardizing your doctor–patient relationship. The practice policy should be clarified. A suitable way to leave this may be to verify your personal policy and leave the form for collection on payment of the required fee.

Question 4

- A patient approaches you to ask you to prescribe the new drug that prevents heart disease. He brings a newspaper cutting with him. How would this make you feel?
- What beliefs may underlie this request?
- Pretend I am that patient. What do you say to me?

Notes

This is not an uncommon scenario for the GP. You need to avoid feeling threatened. Listen to the patient's views and read the article. Understanding his expectations is essential to a good approach. Be ready for a little role play. You may choose to admit your ignorance of the drug, discuss your opinion of the validity of the claims and the way you prescribe medicines e.g. according to guidelines or after proof of value. He should not leave feeling that you are closed to new ideas or only looking after your budget.

Question 5

- A 72 year old lady comes to see you following the death of her husband after 3 weeks on a life support machine. She is certain he would have abhorred this and has come to tell you that she would not want this treatment in similar circumstances. How can you help her?
- Is an advance directive legally binding in the UK?
- What advice would you give to a patient who was thinking about writing one?

Notes

Advance directives (or living wills) ensure the wishes of a competent patient are considered when informed consent can no longer be obtained. Its legal basis is in case law rather than statutory law. It requires a fully informed patient which presents problems with generalized directives as any doubt in intention favours medical intervention. Problems can be minimized by mentioning specific scenarios, reviewing every 5 years and asking the patient to inform her relatives. Clear directives are as legally binding as any current decision by a competent patient. The GPs role is one of advisor and repository. Forms are available from the Euthanasia Society or the Terence Higgins Trust. Further guidance is available from the BMA Ethics Division.

VIVA TABLE 9

Question 1

- How should your consultation skills be checked throughout your career?
- What are your opinions on the use of video to assess consultation skills?
- What makes an "unacceptable GP"?

Notes

The quality of consultations will be a key feature of revalidation. The RCGP's document Good Medical Practice for GPs spells out the attributes of an "excellent GP" and the "unacceptable GP" who has poor communication and professional skills and provides a service which does not incorporate adequate consideration of patients' needs. Patient questionnaires and the use of video in consultations are examples of methods of offering evidence of a high standard of care.

Question 2

- Your practice nurse gives advice which is incorrect. The patient complains. What do you do?
- What is the legal situation – is it your fault?
- What are you going to do to avoid it happening again?

Notes

The practice complaints procedure is required to process the complaint. Practice nurses are professionals with their own insurance but as her employer you must provide adequate training, assessment and review any problems. You must do everything you can to prevent a recurrence and document it. Interview the nurse – there are two sides to every story. Review policies and protocols as necessary. Options for change may include the need for retraining, a revision of policy or creation of new guidelines.

Question 3

- How do you feel when running late?
- What would you do if you consistently felt that way?
- What could help you to organize your day?

Notes

Assess the demand of unscheduled tasks and your behaviour at work. Make sure tasks are prioritised. If it does not need doing don't do it! Allow for delays in your schedule, set realistic targets in manageable chunks, have a systematic approach, learn to say no, delegate where you can and allow time for yourself everyday.

Question 4

- After referring a 20 year old woman for a termination of pregnancy, she requests that you omit the incident from her notes. What issues does this raise for you?

- Will you agree to this request?
- Are there any circumstances in which you would change your mind?

Notes

The answer structure can use a similar approach to an essay answer. Explore the reasons for her request and remind her of total confidentiality. You may or may not agree to her request but you will need to justify your answer in the face of challenge from the examiner. Records may be needed for any future health problems. Incomplete notes may prejudice future management in the event of complications. Your terms of service require you to keep adequate records for the current episode and any future events. They will help if the patient sees a different doctor next time and increase the awareness of possible psychological consequences. They are important in demonstrating a good standard of care. The GMC's Duties of a Doctor include a duty to make and keep proper records. Some omissions at the patient's request may have the intention of misleading for example a life assurance company. Omission in this context could result in an allegation of negligence against the GP.

Question 5

- Do you consider yourself a family doctor?
- Do you feel that home visits are a thing of the past to the modern GP?
- What might be the advantages of home visits?

Notes

Despite the decreasing trend, home visits are not going to go away. There remain requirements for them within our terms of service and there are times when they can help you provide an improved and holistic standard of care. A home visit can allow a functional assessment of the patient at home, help identify hazards, the impact on the household and provide the opportunity to check compliance. They can also reaffirm your role as a family doctor both to you and the patient.

VIVA TABLE 10

Question 1

- What are your feelings on nurse prescribing?
- Have you ever asked a nurse's advice on what to prescribe?
- Do you feel prescribing powers should be extended to pharmacists?

Notes

Prescribing is a reality of modern health care for district nurses and health visitors and expansion of this is advocated by the government. Their experience with dressings and bandaging for example usually exceeds that of doctors. Use of a formulary is important. A distinction between dependent (treatment following doctor diagnosis) and independent (diagnosis and treatment) prescribing is important.

Question 2

- You are called by a patient complaining of severe headache who says she is unable to contact her own GP. What would your feelings be?
- How would you act?
- How close is the relationship between a GP and his colleagues working in the same area?

Notes

Recognize the irritation that this problem will cause and be aware of the possibility of manipulation by the patient. You may be able to contact her own GP or if there is evidence of severe morbidity, call an ambulance on her behalf.
The possibility of a serious diagnosis may compel you to visit. You will need to keep good records in the event of any future complaint.
The least you must do is assess need, act appropriately and inform the patient's GP.

Question 3

- The wife of a patient with terminal cancer comes to you asking for a sick note so that she can stay at home to look after him. She is generally well. What is your reaction?
- Would you not consider such an action dishonest?
- Would you ask for further help on this issue?

Notes

You may feel that documentation of the wife's distress at this unfortunate situation will enable you to sign a sick note. What you write on the sick note can be sufficiently vague e.g. stress-related problem. A less controversial approach would be to volunteer to write to the employer on the patient's behalf if you both felt this would be useful. Alternative strategies may involve offering extra help at home. A sick note renders a patient incapable of work. If no appropriate medical condition exists, supplying a certificate can be seen as clear dishonesty. Your defence union may not have a great deal of sympathy. The "family-friendly policies" of the Employment Relations Bill 1999 may be of use to facilitate unpaid leave.

Question 4

- You diagnose a viral infection in a 9 month old baby. You spend time explaining the diagnosis and prescribe paracetamol but after the mother and baby leave, your receptionist comes to tell you that she wants to see another doctor. How do you feel?
- What options do you have?
- What factors might have influenced her behaviour?

Notes

Initially you may both feel anger. Clearly her expectations have not been met and her anxieties not relieved. Refusal of a second opinion will not help anybody. You may decide to offer to see her again if this is acceptable. A later appointment with a colleague will at least avoid an attendance at A&E or a possible night visit. Discussion with the relevant partner will make sure you are all singing from the same hymn sheet while allowing for the possibility of a diagnosis overlooked.

Question 5

- How important is a good receptionist to a practice?
- Do they have a role in triage?
- How would we audit the work they do?

Notes

A good receptionist is a priceless asset. They are the interface and first point of contact with the practice either by phone or in person. A good use of common sense allows them to help patients prioritize their appointments. A system needs to be in place which avoids them making decisions outside of their competence. Doctors need to be sufficiently accessible during surgery time to deal with any problems.
The GP must decide what tasks to delegate to the receptionist and provide adequate training.

FURTHER VIVA PRACTICE

VIVA 1

- Do you ever prescribe antibiotics over the phone?
- What are the dangers of doing so?
- How could you justify doing so in the event of a significant reaction?

- How will you manage family life and life as a GP?
- What difficulties does having doctors as patients present?
- How would you recognize burnout in a colleague?

- Tell me about something interesting you have read recently.
- How do you know what you don't know?
- Can a team approach to education benefit the whole practice?

- Do you think the NHS should finance nicotine replacement?
- When do you use private prescriptions?
- When do you refer patients privately?

- You overhear your staff openly discussing the case history of one of your patients. How do you react?
- What efforts do you take in your practice to ensure confidentiality?
- What do you understand by the term "informed consent"?

VIVA 2

- Your partner refuses to give contraception or advice about terminations. What problems does this raise for the practice?
- You receive a complaint from a patient's participation group. How do you react?
- Do your personal feelings affect the way you perform as a GP?

- How can Primary Care Education Centres benefit the GP?
- How can we make Continuing Medical Education as effective as possible?
- How might you measure your professional achievements and shortcomings in 5 years' time?

- Do you think there is any value in self-help groups?
- How would you react if you were asked to talk to the multiple sclerosis self-help group about the latest treatments?
- Do you ever suggest that your patients contact one of these groups? Why/Why not?

- A wife asks you not to tell her husband that he has cancer. What is your duty to your patient?
- You visit her 2 weeks after the death of her husband. What do you wish to discuss?
- What is your plan for follow-up?

- What difficulties do you think your practice nurse faces in her job?
- What about the receptionists?
- How can you make sure these problems are recognized and try to improve communication in the primary health care team?

VIVA 3

- What do patients want from their GP?
- Are you aware of any legislation that has increased patients' expectations?
- A patient demands a visit for a minor complaint. What do you do?

- What is PACT (prescribing analyses and cost) data?
- What value is it and what are its limitations?
- How do you see its uses expanding?

- A patient has borrowed some omeprazole from a friend and found it has worked wonders for his dyspepsia. He comes to you for some more. How do you start the consultation?
- What are the issues this raises?
- How do you educate patients?

- How has computerization affected general practice?
- How does it affect your day to day work?
- How would you like to see it developing in the future?

- When you are a principal what piece of equipment would you most like that you have not got now?
- How will you convince me as your partner?
- How might it be financed?

VIVA 4

- Your next patient has been asked to attend after a severely dyskaryotic cervical smear result taken 8 months ago appears not to have been acted on. How would you have planned for this?
- How are you going to make sure she takes in what you say?
- How do you end the consultation?

- Your next patient has missed their last four appointments. What problems does this cause?
- How will you try to help the problem?
- What could the practice do as a group?

- You see one of your patients who has just been diagnosed with epilepsy driving down the high street. What action do you take?
- How can you progress without breaking confidentiality?
- Can you enlist anybody to help you?

- Why do people consult alternative practitioners?
- A patient of yours asks you to recommend an osteopath. How do you reply?
- Where would you advise a patient to find out more about homeopathy?

- What piece of litigation would you like to see to improve the standard of health care?
- What if there was no extra money available?
- How would you help promote a Give Up Smoking campaign?

VIVA 5

- Tell me about a paper that has influenced your medical practice.
- What book would you recommend to a friend who wishes to pass the MRCGP?
- What non-medical book would you recommend?

- Would you ever break the confidence of a patient?
- Would you inform a surgeon about the HIV positive status of a patient you are referring for hernia repair?
- How would you go about this?

- What do you think varies the rate of referral between GPs and between practices?
- Do you think this could be used as a marker of quality?
- Would you be happy to have your practice data published in a league table?

- During a minor surgery procedure, you discover you have accidentally injected a small amount of lignocaine with adrenaline into someone's finger. Do you admit the mistake?
- The amount is only very small and there will almost certainly be no after-effects. How do you react now?
- What are the consequences of each approach? Is total honesty always the best policy?

- You see one of your partners at a restaurant with the practice manager. Would this concern you?
- Would you tell anyone?
- How might it compromise the team?

VIVA 6

- What is happening to the profession at the moment?
- What form do you think reaccreditation should take?
- What is mentorship?

- An 80 year old lady with severe arthritis asks you to explain how many of her phenobarbitone tablets she would need to take in order to kill herself. What issues does this raise?
- How would you decide what to tell her?
- She is found the next day dead in bed. The family are against a post-mortem. Would you inform the coroner?

- You are telephoned by the police to attend a violent patient. How will you prepare yourself?
- What issues does this raise?
- What measures could you take to improve your safety at work?

- A lady who is the sole carer of her increasingly dependent mother says she is getting too much for her and tells you "you'll have to put her in a home!" How does this make you feel?
- Who is the patient here?
- What options can you suggest?

- You receive a call from a nursing home regarding a resident who has fallen out of bed and has pain in the leg. She asks if you want to visit or if she should just call an ambulance. What is your reply?
- Have you ever told a patient to go to A&E when you could have seen them yourself?
- How should a GP communicate with an A&E department?

VIVA 7

- You notice that one of your partners is becoming increasingly irritable and smelling of alcohol. You know that he is going through a divorce. How will you react?
- You feel that he is making mistakes. What will you do?
- Your health visitor comes to tell you that several mothers have complained he has alcohol on his breath. How do you respond?

- A patient asks for a repeat prescription for his father who is now living in Pakistan. Where does your clinical responsibility lie?
- Where would you go for further advice?
- What other options could you suggest?

- A newly registered patient comes to tell you he takes 100 mg of diazepam a day. He demands a 4 week supply. What do you say to him?
- How can you get more information?
- How will you formulate a plan of care?

- What do your patients need from you?
- What is a patient-centred consultation?
- What problems does this approach present?

- How would you assess your consultation skill?
- What does it mean to be a gatekeeper?
- How do financial limitations affect your day-to-day practice?

VIVA 8

- It could be said that modern society is becoming deskilled at handling minor illness. Why do you think that is?
- What would you do to change it?
- How would you aim to redefine the term "minor illness"?

- How would you explain the diagnosis of asthma to the mother of a wheezing 3 year old?
- How do you find her level of understanding?
- How would you close the consultation?

- How well do you work personally in a team?
- What are your strengths and weaknesses?
- What situations in practice challenge your personal values and effectiveness as a doctor?

- How would you decide whether your primary care team is effective or not?
- What outcome measures would you like to target?
- Patients are crucial to our effectiveness, why do we not include them in our teams?

- How is the profession policed?
- How will you cope with pressure of a complaint?
- How will you approach the future in the face in increasing litigation and complaints?

VIVA 9

- Do you think your practice should possess an ECG machine?
- Would you carry a defibrillator?
- What are the differences between rural and urban practice?

- What audit have you been involved in recently?
- Did it change practice?
- How would you choose an area of your practice to audit?

- How can you assess the needs of your patients and respond to them?
- What would you do if someone asked you to prescribe a tonic?
- Would you prescribe vitamin supplements?

- What is telemedicine?
- How can technological developments improve patient care?
- How do you see the work of a GP in 20 years' time?

- You receive a telephone call from the local pharmacist saying that the prescription you gave to your previous patient has the wrong name on it. What is your reaction?
- How can you avoid such mistakes happpening?
- How might this affect your relationship with staff and patients?

VIVA 10

- A young female patient of yours dies from an overdose of an antidepressant you prescribed. Describe your thoughts and feelings about this.
- Do you know any evidence that shows whether the availability of prescribed drugs is related to their use in self-poisoning?
- How would you avoid this happening again?

- You are asked to give a talk on improving health care in areas of poverty. How would you prepare?
- What measures of deprivation do you know?
- How would you increase your awareness of your patients' problems throughout your career?

- You receive a request for slimming tablets. What is your reply?
- Do you ever prescribe medicines without clear evidence of their benefit?
- Do you think that you should aim to avoid this as your career continues?

- You meet one of your patients in the street. Would you ask her how she is?
- Are you prepared to be always on duty?
- How might your answer affect you and those around you?

- You become aware that an HIV patient of yours has not informed his partner of his status. Where does your duty of care lie?
- His partner comes to see you for advice on starting a family. What will you say?
- Would you be happy to breach confidentiality?

FURTHER DILEMMAS

1. A woman comes to you requesting a paternity test for her 4 year old son. She had an affair shortly before pregnancy and would like to be sure her husband is the father.
2. A practice nurse asks you to prescribe antibiotics for a patient you have not seen.
3. A man is visited in the night by the deputizing service. He is given amoxycillin despite telling the doctor about his penicillin allergy. He sees you the following day with a severe rash consistent with his allergy.
4. A terminally ill patient requests help to end his life.
5. A consultant asks you to prescribe a treatment that he cannot afford to on his budget.
6. What sort of procedures do you think should be outside the scope of the NHS due to budget constraints?
7. A 40 year old patient asks you to refer him for reversal of vasectomy.
8. A 20 year old epileptic comes to see you with the good news that she has had no fits for a year and has stopped her medication.
9. A patient asks you to release his notes to a nearby private GP.
10. A patient requests a second opinion that you do not believe to be necessary.
11. A patient is abusive to staff.
12. You receive a request to visit a lady who has suffered miscarriage 2 days previously.
13. A mother phones after finding her 5 month old baby apparently dead.
14. You receive two complaints regarding one of your receptionists.
15. A patient with chronic renal failure tells you he is no longer willing to submit to dialysis.
16. A partner suggests that you should provide an evening clinic for the many people who find difficulty getting to the surgery.
17. A 16 year old girl in the middle of her GCSEs presents with a sore throat requesting antibiotics.
18. A moribund terminal care patient looked after by visits from yourself and a district nurse lives with a caring wife and son. On attending, you find his syringe driver which had nearly 20 hours to run is empty. The patient is dead.
19. A mother of a 14 year old girl you have recently started on the pill asks to see her daughter's notes.

20. A patient has returned from a specialist with a diagnosis of retrobulbar neuritis. He comes to you to clarify things. He was told he has some inflammation of the nerve behind the eye which has settled now.
21. You are asked to provide a sip-feed (a borderline substance) for a nursing home resident.
22. A 26 year old lawyer considering pregnancy comes to request immunization against chickenpox.
23. A private GP clinic is set up locally offering 24 hour care including telephone consults and home visits.
24. You see a man whom you have certified sick with back pain for the past 3 months building a garden wall.
25. A patient with rhinitis has not responded to treatment. She requests referral to a distant clinical ecology consultant for food allergy.

APPENDIX 1

APPRAISING THE ARTICLE

This checklist of questions to ask yourself is a useful way of analysing clinical papers. Comments can be made on each point in a positive or negative vein to create an answer with structure.

Introduction

- Are the aims stated clearly?
- Does the study match up to the aims?
- Is literature reviewed to place the article in perspective?

Method

- What is the study design?
  e.g. qualitative/quantitative, observational/experimental, retrospective/prospective, longitudinal/cross-sectional
- Is the study design appropriate?
- Is there a gold standard for comparison?
- Do the instruments have validity in practice?
  e.g. are questions designed to avoid bias and ambiguity?

Are the inclusion (and exclusion) criteria clear?

- Is the population representative of a GP population?
  e.g. similar age–sex distribution
- Is power calculated and the sample size sufficient for significance?
- Is the sample unchanged during the study?
- Are the controls appropriate?
- Is the method of randomization described sufficiently to allow reproduction of the experiment? Is it fair?
- Is the time-span defined and appropriate?
- Is treatment clear?
- Are outcome criteria clearly defined?
- Are the end-points soft or hard? Are they appropriate?
- Are all relevant outcomes included?
- Is it truly blind to patients and clinicians?

Results

- Is the response rate reasonable? (above 70% is good)
- Is follow-up adequate?
- Are all subjects accounted for?
  e.g. is analysis by intention to treat and are drop-outs adequately explained?
- Are data represented honestly with tables and graphs?
- Are statistics appropriate for all the findings?
- Do they include confidence limits and p values?

Discussion

- Were the aims met?
- Is there an objective discussion of limitations and applicability?
- Are conclusions justified and speculations realistic?
- Is there comparison with previous research?

Overall

- Was the study clear, valid, ethical and worthwhile?
- Are the conclusions affordable, available and sensible enough to bring about a change in practice?

APPENDIX 2

KEY TO EVIDENCE STATEMENTS AND GRADES OF RECOMMENDATIONS

The hierarchy of evidence can be described as follows

Statements of evidence

Ia Evidence obtained from meta-analysis of randomized controlled trials.
Ib Evidence obtained from at least one randomized controlled trial.
IIa Evidence obtained from at least one well-designed controlled study without randomization.
IIb Evidence obtained from at least one other type of well-designed quasi-experimental study.
III Evidence obtained from well-designed non-experimental descriptive studies, such as comparative studies, correlation studies and case studies.
IV Evidence obtained from expert committee reports or opinions and/or clinical experiences of respected authorities.

Grades of recommendations

A. Requires at least one randomized controlled trial as part of a body of literature of overall good quality and consistency addressing the specific recommendation.
(Evidence levels Ia, Ib)
B. Requires the availability of well conducted clinical studies but no randomized clinical trials on the topic of recommendation.
(Evidence levels IIa, IIb, III)
C. Requires evidence obtained from expert committee reports or opinions and/or clinical experiences of respected authorities. Indicates an absence of directly applicable clinical studies of good quality.
(Evidence level IV)

APPENDIX 3

HOT TOPICS

This list may be a useful starting point to identify topics of value but does not claim to be comprehensive. Keeping up-to-date is essential to identify current key topics in general practice.

Non-clinical areas

- Complaints procedures.
- Confidentiality and HIV.
- Development of screening and health promotion.
- Informed consent.
- Lifelong learning.
- New initiatives in primary care.
- Open access to services.
- Out of hours work patterns.
- Poverty and deprivation.
- Practice formularies.
- Prescribing data.
- Reaccreditation.
- Role of counsellors.
- Stress and burnout.

Clinical management

- Alcohol and drugs.
- Antiplatelet/anticoagulation and monitoring in practice.
- Asthma – self-management and guidelines.
- Breast disease and screening protocols.
- Care of the elderly.
- Chlamydia.
- Depression and the role of the GP.
- Diabetic care.
- Domiciliary thrombolysis.
- Heart failure and the use of ACE inhibitors.
- *Helicobacter pylori.*
- HRT and evidence of benefit.
- Hyperlipidaemia and current recommendations.
- Hypertension and guidelines.
- Intrapartum care at home.
- Minor illness including sore throat and otitis media.
- Screening for prostate and bowel cancer.
- Smoking cessation.
- Stroke management.

INDEX

Abortion, 101
ACE inhibitors 10,11
Adolescents 6,32
Advance directives 126
AIDS 57
Alcohol 8,41,99
Alzheimer's disease 70
Antibiotics 58
Appraisal 119
Aspirin 10,39
Asthma 50

Back pain 55
Bad news 63,78,112
Benzodiazepines 22
Bereavement 78
Beta-blockers 5,10,80
Boxing 111
Burnout 62,111

Cancer colon prostate 42,43
Cannabis 125
CHD 8, 10,20,23
Chlamydia 76,101
Cholesterol 10,23
Chronic fatigue syndrome 87
Clinical governance 120
Complaints 121,127
Compliance 29,80
Consultations, longer 66
Contraception 33,60
COPD 28
Cosmetic surgery 113
Counselling 36,44
Cultural differences 7

Dementia 70
Depression 34,44
Deputising service 82
Diabetes 18, 20
Diuretics 5
Drug misuse 3,16
Dyspepsia 89

Echocardiography 11
Elderly 5,84,85
Epilepsy 93
Exercise 8,13

Folate 27

Gifts 46
Gillick competence 6,32,118
Grief 78
Guidelines 28,40,50,52,55,58,80

Heart failure 11
Heartsinks 123
Helicobacter 89
Heroin 3,16
HIV 3,57
Home birth 104
Hormone replacement 25
Hyperactivity 47
Hyperlipidaemia 23
Hypertension 5,20,80

Insomnia 22
Intrapartum care 104

Lifestyles 8,13
Living wills 126

ME 87
MI 10
Models of care 93
Morbidity registers 112

Nicotine replacement 8,30
Nursing homes 84

Obesity 8,
Osteoporosis 13,26
Out of hours 82
Overdose 91

PACT data 14,120
Palliative care 48,
Patient-held records 111
Pharmacists 69
Practice nurses 124,127
Preconceptual counselling 27
Prescribing 14,129
Prescribing formulary 45,69

Prevention 8,13,23, 75,95
Prostate 43

Quality 120

Rationing 73
Receptionists 130
Research 119
Rehabilitation cardiac 8

stroke 38
Schizophrenia 96
Screening 42,75,76,85
Self-harm 91
Sigmoidoscopy 42
Smoking 8,27,28,30
Soiling 54
Sore throat 58
Spirometry 28
Statins 10
Stress 8,49,62,111
Stroke 38

Telephone consultations 64
Termination of pregnancy 101
Thrombolysis 37
Triage 64,130

UTI 40

Vaccination 98
Violence 67,71,102

Wilson's criteria 42,76
Walk-in centres 106